Big Data Analysis for Bioinformatics and Biomedical Discoveries

CHAPMAN & HALL/CRC
Mathematical and Computational Biology Series

Aims and scope:

This series aims to capture new developments and summarize what is known over the entire spectrum of mathematical and computational biology and medicine. It seeks to encourage the integration of mathematical, statistical, and computational methods into biology by publishing a broad range of textbooks, reference works, and handbooks. The titles included in the series are meant to appeal to students, researchers, and professionals in the mathematical, statistical and computational sciences, fundamental biology and bioengineering, as well as interdisciplinary researchers involved in the field. The inclusion of concrete examples and applications, and programming techniques and examples, is highly encouraged.

Series Editors

N. F. Britton
Department of Mathematical Sciences
University of Bath

Xihong Lin
Department of Biostatistics
Harvard University

Nicola Mulder
University of Cape Town
South Africa

Maria Victoria Schneider
European Bioinformatics Institute

Mona Singh
Department of Computer Science
Princeton University

Anna Tramontano
Department of Physics
University of Rome La Sapienza

Proposals for the series should be submitted to one of the series editors above or directly to:
CRC Press, Taylor & Francis Group
3 Park Square, Milton Park
Abingdon, Oxfordshire OX14 4RN
UK

Published Titles

Published Titles (continued)

Chapman & Hall/CRC Mathematical and Computational Biology Series

Big Data Analysis for Bioinformatics and Biomedical Discoveries

Edited by
Shui Qing Ye

CRC Press
Taylor & Francis Group
Boca Raton   London   New York

CRC Press is an imprint of the
Taylor & Francis Group, an **informa** business
A CHAPMAN & HALL BOOK

Cover Credit:

Foreground image: Zhang LQ, Adyshev DM, Singleton P, Li H, Cepeda J, Huang SY, Zou X, Verin AD, Tu J, Garcia JG, Ye SQ. Interactions between PBEF and oxidative stress proteins - A potential new mechanism underlying PBEF in the pathogenesis of acute lung injury. FEBS Lett. 2008; 582(13):1802-8

Background image: Simon B, Easley RB, Gregoryov D, Ma SF, Ye SQ, Lavoie T, Garcia JGN. Microarray analysis of regional cellular responses to local mechanical stress in experimental acute lung injury. Am J Physiol Lung Cell Mol Physiol. 2006; 291(5):L851-61

CRC Press
Taylor & Francis Group
6000 Broken Sound Parkway NW, Suite 300
Boca Raton, FL 33487-2742

First issued in paperback 2021

© 2016 by Taylor & Francis Group, LLC
CRC Press is an imprint of Taylor & Francis Group, an Informa business

No claim to original U.S. Government works

Version Date: 20151112

ISBN-13: 978-0-367-78327-3 (pbk)
ISBN-13: 978-1-4987-2452-4 (hbk)

Contents

Preface

W E ARE ENTERING AN era of *Big Data*. Big Data offer both unprecedented opportunities and overwhelming challenges. This book is intended to provide biologists, biomedical scientists, bioinformaticians, computer data analysts, and other interested readers with a pragmatic blueprint to the nuts and bolts of Big Data so they more quickly, easily, and effectively harness the power of Big Data in their ground-breaking biological discoveries, translational medical researches, and personalized genomic medicine.

Big Data refers to increasingly larger, more diverse, and more complex data sets that challenge the abilities of traditionally or most commonly used approaches to access, manage, and analyze data effectively. The monumental completion of human genome sequencing ignited the generation of big biomedical data. With the advent of ever-evolving, cutting-edge, high-throughput omic technologies, we are facing an explosive growth in the volume of biological and biomedical data. For example, Gene Expression Omnibus (http://www.ncbi.nlm.nih.gov/geo/) holds 3,848 data sets of transcriptome repositories derived from 1,423,663 samples, as of June 9, 2015. Big biomedical data come from government-sponsored projects such as the 1000 Genomes Project (http://www.1000genomes.org/), international consortia such as the ENCODE Project (http://www.genome.gov/encode/), millions of individual investigator-initiated research projects, and vast pharmaceutical R&D projects. Data management can become a very complex process, especially when large volumes of data come from multiple sources and diverse types, such as images, molecules, phenotypes, and electronic medical records. These data need to be linked, connected, and correlated, which will enable researchers to grasp the information that is supposed to be conveyed by these data. It is evident that these Big Data with high-volume, high-velocity, and high-variety information provide us both tremendous opportunities and compelling challenges. By leveraging the diversity of available molecular and clinical Big Data, biomedical scientists can now gain new unifying global biological insights into human physiology and the molecular pathogenesis of various human diseases or conditions at an unprecedented scale and speed; they can also identify new potential candidate molecules that have a high probability of being successfully developed into drugs that act on biological targets safely and effectively. On the other hand, major challenges in using biomedical Big Data are very real, such as how to have a knack for some Big Data analysis software tools, how to analyze and interpret various next-generation DNA sequencing data, and how to standardize and integrate various big biomedical data to make global, novel, objective, and data-driven discoveries. Users of Big Data can be easily "lost in the sheer volume of numbers."

The objective of this book is in part to contribute to the NIH Big Data to Knowledge (BD2K) (http://bd2k.nih.gov/) initiative and enable biomedical scientists to capitalize on the Big Data being generated in the omic age; this goal may be accomplished by enhancing the computational and quantitative skills of biomedical researchers and by increasing the number of computationally and quantitatively skilled biomedical trainees.

This book covers many important topics of Big Data analyses in bioinformatics for biomedical discoveries. Section I introduces commonly used tools and software for Big Data analyses, with chapters on Linux for Big Data analysis, Python for Big Data analysis, and the R project for Big Data computing. Section II focuses on next-generation DNA sequencing data analyses, with chapters on whole-genome-seq data analysis, RNA-seq data analysis, microbiome-seq data analysis, miRNA-seq data analysis, methylome-seq data analysis, and ChIP-seq data analysis. Section III discusses comprehensive Big Data analyses of several major areas, with chapters on integrating omics data with Big Data analysis, pharmacogenetics and genomics, exploring de-identified electronic health record data with i2b2, Big Data and drug discovery, literature-based knowledge discovery, and mitigating high dimensionality in Big Data analysis. All chapters in this book are organized in a consistent and easily understandable format. Each chapter begins with a theoretical introduction to the subject matter of the chapter, which is followed by its exemplar applications and data analysis principles, followed in turn by a step-by-step tutorial to help readers to obtain a good theoretical understanding and to master related practical applications. Experts in their respective fields have contributed to this book, in common and plain English. Complex mathematical deductions and jargon have been avoided or reduced to a minimum. Even a novice, with little knowledge of computers, can learn Big Data analysis from this book without difficulty. At the end of each chapter, several original and authoritative references have been provided, so that more experienced readers may explore the subject in depth. The intended readership of this book comprises biologists and biomedical scientists; computer specialists may find it helpful as well.

I hope this book will help readers demystify, humanize, and foster their biomedical and biological Big Data analyses. I welcome constructive criticism and suggestions for improvement so that they may be incorporated in a subsequent edition.

Shui Qing Ye
University of Missouri at Kansas City

MATLAB® is a registered trademark of The MathWorks, Inc. For product information, please contact:

The MathWorks, Inc.
3 Apple Hill Drive
Natick, MA 01760-2098 USA
Tel: 508-647-7000
Fax: 508-647-7001
E-mail: info@mathworks.com
Web: www.mathworks.com

Acknowledgments

I SINCERELY APPRECIATE DR. SUNIL NAIR, a visionary publisher from CRC Press/Taylor & Francis Group, for granting us the opportunity to contribute this book. I also thank Jill J. Jurgensen, senior project coordinator; Alex Edwards, editorial assistant; and Todd Perry, project editor, for their helpful guidance, genial support, and patient nudge along the way of our writing and publishing process.

I thank all contributing authors for committing their precious time and efforts to pen their valuable chapters and for their gracious tolerance to my haggling over revisions and deadlines. I am particularly grateful to my colleagues, Dr. Daniel P. Heruth and Dr. Min Xiong, who have not only contributed several chapters but also carefully double checked all next-generation DNA sequencing data analysis pipelines and other tutorial steps presented in the tutorial sections of all chapters.

Finally, I am deeply indebted to my wife, Li Qin Zhang, for standing beside me throughout my career and editing this book. She has not only contributed chapters to this book but also shouldered most responsibilities of gourmet cooking, cleaning, washing, and various household chores while I have been working and writing on weekends, nights, and other times inconvenient to my family. I have also relished the understanding, support, and encouragement of my lovely daughter, Yu Min Ye, who is also a writer, during this endeavor.

Editor

Shui Qing Ye, MD, PhD, is the William R. Brown/Missouri endowed chair in medical genetics and molecular medicine and a tenured full professor in biomedical and health informatics and pediatrics at the University of Missouri–Kansas City, Missouri. He is also the director in the Division of Experimental and Translational Genetics, Department of Pediatrics, and director in the Core of Omic Research at The Children's Mercy Hospital. Dr. Ye completed his medical education from Wuhan University School of Medicine, Wuhan, China, and earned his PhD from the University of Chicago Pritzker School of Medicine, Chicago, Illinois. Dr. Ye's academic career has evolved from an assistant professorship at Johns Hopkins University, Baltimore, Maryland, followed by an associate professorship at the University of Chicago to a tenured full professorship at the University of Missouri at Columbia and his current positions.

Dr. Ye has been engaged in biomedical research for more than 30 years; he has experience as a principal investigator in the NIH-funded RO1 or pharmaceutical company–sponsored research projects as well as a co-investigator in the NIH-funded RO1, Specialized Centers of Clinically Oriented Research (SCCOR), Program Project Grant (PPG), and private foundation fundings. He has served in grant review panels or study sections of the National Heart, Lung, Blood Institute (NHLBI)/National Institutes of Health (NIH), Department of Defense, and American Heart Association. He is currently a member in the American Association for the Advancement of Science, American Heart Association, and American Thoracic Society. Dr. Ye has published more than 170 peer-reviewed research articles, abstracts, reviews, book chapters, and he has participated in the peer review activity for a number of scientific journals.

Dr. Ye is keen on applying high-throughput genomic and transcriptomic approaches, or *Big Data*, in his biomedical research. Using direct DNA sequencing to identify single-nucleotide polymorphisms in patient

DNA samples, his lab was the first to report a *susceptible haplotype* and a *protective haplotype* in the human pre-B-cell colony-enhancing factor gene promoter to be associated with acute respiratory distress syndrome. Through a DNA microarray to detect differentially expressed genes, Dr. Ye's lab discovered that the pre-B-cell colony-enhancing factor gene was highly upregulated as a biomarker in acute respiratory distress syndrome. Dr. Ye had previously served as the director, Gene Expression Profiling Core, at the Center of Translational Respiratory Medicine in Johns Hopkins University School of Medicine and the director, Molecular Resource Core, in an NIH-funded Program Project Grant on Lung Endothelial Pathobiology at the University of Chicago Pritzker School of Medicine. He is currently directing the Core of Omic Research at The Children's Mercy Hospital, University of Missouri–Kansas City, which has conducted exome-seq, RNA-seq, miRNA-seq, and microbiome-seq using state-of-the-art next-generation DNA sequencing technologies. The Core is continuously expanding its scope of service on omic research. Dr. Ye, as the editor, has published a book entitled *Bioinformatics: A Practical Approach* (CRC Press/Taylor & Francis Group, New York). One of Dr. Ye's current and growing research interests is the application of translational bioinformatics to leverage *Big Data* to make biological discoveries and gain new unifying global biological insights, which may lead to the development of new diagnostic and therapeutic targets for human diseases.

Contributors

Chengpeng Bi
Division of Clinical Pharmacology,
    Toxicology, and Therapeutic
    Innovations
The Children's Mercy Hospital
University of Missouri-Kansas
    City School of Medicine
Kansas City, Missouri

Guang-Liang Bi
Department of Neonatology
Nanfang Hospital, Southern
    Medical University
Guangzhou, China

Larisa H. Cavallari
Department of Pharmacotherapy
    and Translational Research
Center for Pharmacogenomics
University of Florida
Gainesville, Florida

Deendayal Dinakarpandian
Department of Computer
    Science and Electrical
    Engineering
University of Missouri-Kansas
    City School of Computing and
    Engineering
Kansas City, Missouri

Andrea Gaedigk
Division of Clinical Pharmacology,
    Toxicology & Therapeutic
    Innovation
Children's Mercy Kansas City
and
Department of Pediatrics
University of Missouri-Kansas
    City School of Medicine
Kansas City, Missouri

Dmitry N. Grigoryev
Laboratory of Translational
    Studies and Personalized
    Medicine
Moscow Institute of Physics and
    Technology
Dolgoprudny, Moscow, Russia

Daniel P. Heruth
Division of Experimental and
    Translational Genetics
Children's Mercy Hospitals and
    Clinics
and
University of Missouri-Kansas
    City School of Medicine
Kansas City, Missouri

Mark Hoffman
Department of Biomedical
  and Health Informatics and
  Department of Pediatrics
Center for Health Insights
University of Missouri-Kansas
  City School of Medicine
Kansas City, Missouri

Xun Jiang
Department of Pediatrics, Tangdu
  Hospital
The Fourth Military Medical
  University
Xi'an, Shaanxi, China

Ding-You Li
Division of Gastroenterology
Children's Mercy Hospitals and
  Clinics
and
University of Missouri-Kansas
  City School of Medicine
Kansas City, Missouri

Hongfang Liu
Biomedical Statistics and
  Informatics
Mayo Clinic
Rochester, Minnesota

Majid Rastegar-Mojarad
Biomedical Statistics and
  Informatics
Mayo Clinic
Rochester, Minnesota

Katrin Sangkuhl
Department of Genetics
Stanford University
Stanford, California

Stephen D. Simon
Department of Biomedical
  and Health Informatics
University of Missouri-
  Kansas City School of Medicine
Kansas City, Missouri

D. Andrew Skaff
Division of Molecular Biology and
  Biochemistry
University of Missouri-Kansas
  City School of Biological
  Sciences
Kansas City, Missouri

Jiancheng Tu
Department of Clinical
  Laboratory Medicine
Zhongnan Hospital
Wuhan University School of
  Medicine
Wuhan, China

Gerald J. Wyckoff
Division of Molecular Biology
  and Biochemistry
University of Missouri-Kansas
  City School of Biological
  Sciences
Kansas City, Missouri

Min Xiong
Division of Experimental and
  Translational Genetics
Children's Mercy Hospitals and
  Clinics
and
University of Missouri-Kansas
  City School of Medicine
Kansas City, Missouri

Li Qin Zhang
Division of Experimental and
  Translational Genetics
Children's Mercy Hospitals and
  Clinics
and
University of Missouri-Kansas
  City School of Medicine
Kansas City, Missouri

I

Commonly Used Tools
for Big Data Analysis

## CHAPTER 1

Linux for Big Data Analysis

Shui Qing Ye and Ding-you Li

CONTENTS

1.1 INTRODUCTION

As biological data sets have grown larger and biological problems have become more complex, the requirements for computing power have also grown. Computers that can provide this power generally use the Linux/ Unix operating system. Linux was developed by Linus Benedict Torvalds when he was a student in the University of Helsinki, Finland, in early 1990s. Linux is a modular Unix-like computer operating system assembled under the model of free and open-source software development and distribution. It is the leading operating system on servers and other big iron systems such as mainframe computers and supercomputers. Compared to the Windows operating system, Linux has the following advantages:

1. *Low cost*: You don't need to spend time and money to obtain licenses since Linux and much of its software come with the GNU General Public License. GNU is a recursive acronym for *GNU's Not Unix!*. Additionally, there are large software repositories from which you can freely download for almost any task you can think of.

2. *Stability*: Linux doesn't need to be rebooted periodically to maintain performance levels. It doesn't freeze up or slow down over time due to memory leaks. Continuous uptime of hundreds of days (up to a year or more) are not uncommon.

3. *Performance*: Linux provides persistent high performance on workstations and on networks. It can handle unusually large numbers of users simultaneously and can make old computers sufficiently responsive to be useful again.

4. *Network friendliness*: Linux has been continuously developed by a group of programmers over the Internet and has therefore strong

support for network functionality; client and server systems can be easily set up on any computer running Linux. It can perform tasks such as network backups faster and more reliably than alternative systems.

5. *Flexibility*: Linux can be used for high-performance server applications, desktop applications, and embedded systems. You can save disk space by only installing the components needed for a particular use. You can restrict the use of specific computers by installing, for example, only selected office applications instead of the whole suite.

6. *Compatibility*: It runs all common Unix software packages and can process all common file formats.

7. *Choice*: The large number of Linux distributions gives you a choice. Each distribution is developed and supported by a different organization. You can pick the one you like the best; the core functionalities are the same and most software runs on most distributions.

8. *Fast and easy installation*: Most Linux distributions come with user-friendly installation and setup programs. Popular Linux distributions come with tools that make installation of additional software very user friendly as well.

9. *Full use of hard disk*: Linux continues to work well even when the hard disk is almost full.

10. *Multitasking*: Linux is designed to do many things at the same time; for example, a large printing job in the background won't slow down your other work.

11. *Security*: Linux is one of the most secure operating systems. Attributes such as *fireWalls* or flexible file access permission systems prevent access by unwanted visitors or viruses. Linux users have options to select and safely download software, free of charge, from online repositories containing thousands of high-quality packages. No purchase transactions requiring credit card numbers or other sensitive personal information are necessary.

12. *Open Source*: If you develop a software that requires knowledge or modification of the operating system code, Linux's source code is at your fingertips. Most Linux applications are open-source as well.

1.2 RUNNING BASIC LINUX COMMANDS

There are two modes for users to interact with the computer: command-line interface (CLI) and graphical user interface (GUI). CLI is a means of interacting with a computer program where the user issues commands to the program in the form of successive lines of text. GUI allows the use of icons or other visual indicators to interact with a computer program, usually through a mouse and a keyboard. GUI operating systems such as Window are much easier to learn and use because commands do not need to be memorized. Additionally, users do not need to know any programming languages. However, CLI systems such as Linux give the user more control and options. CLIs are often preferred by most advanced computer users. Programs with CLIs are generally easier to automate via scripting, called as *pipeline*. Thus, Linux is emerging as a powerhouse for Big Data analysis. It is advisable to master some basic CLIs necessary to efficiently perform the analysis of Big Data such as next-generation DNA sequence data.

1.2.1 Remote Login to Linux Using Secure Shell

Secure shell (SSH) is a cryptographic network protocol for secure data communication, remote command-line login, remote command execution, and other secure network services between two networked computers. It connects, via a secure channel over an insecure network, a server and a client running SSH server and SSH client programs, respectively. Remote login to Linux compute server needs to use an SSH. Here, we use PuTTY as an SSH client example. PuTTY was developed originally by Simon Tatham for the Windows platform. PuTTY is an open-source software that is available with source code and is developed and supported by a group of volunteers. PuTTY can be freely and easily downloaded from the site (http://www.putty.org/) and installed by following the online instructions. Figure 1.1a displays the starting portal of a PuTTY SSH. When you input an IP address under Host Name (or IP address) such as 10.250.20.231, select Protocol SSH, and then click Open; a login screen will appear. After successful login, you are at the input prompt $ as shown in Figure 1.1b and the shell is ready to receive proper command or execute a script.

1.2.2 Basic Linux Commands

Table 1.1 lists most common basic commands used in Linux operation. To learn more about the various commands, one can type **man** program

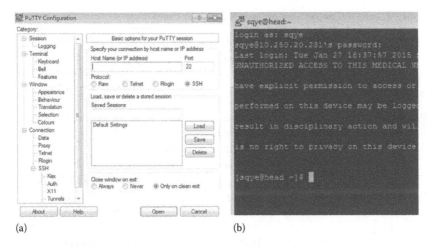

(a)   (b)

FIGURE 1.1   Screenshots of a PuTTy confirmation (a) and a valid login to Linux (b).

TABLE 1.1   Common Basic Linux Commands

Category	Command	Description	Example
File administration	ls	List files	ls -al, list all file in detail
	cp	Copy source file to target file	cp myfile yourfile
	rm	Remove files or directories (rmdir or rm -r)	rm accounts.txt, to remove the file "accounts.txt" in the current directory
	cd	Change current directory	cd., to move to the parent directory of the current directory
	mkdir	Create a new directory	mkdir mydir, to create a new directory called mydir
	gzip/gunzip	Compress/uncompress the contents of files	gzip .swp, to compress the file .swp
Access file contents	cat	Display the full contents of a file	cat Mary.py, to display the full content of the file "Mary.py"
	Less/more	Browse the contents of the specified file	less huge-log-file.log, to browse the content of huge-log-file.log
	Tail/head	Display the last or the first 10 lines of a file by default	tail -n N filename.txt, to display N number of lines from the file named filename.txt
	find	Find files	find ~ -size -100M, To find files smaller than 100M

(*Continued*)

TABLE 1.1 (*CONTINUED*) Common Basic Linux Commands

Category	Command	Description	Example
	grep	Search for a specific string in the specified file	grep "this" demo_file, to search "this" containing sentences from the "demo_file"
Processes	top	Provide an ongoing look at processor activity in real time	top –s, to work in secure mode
	kill	Shut down a process	kill -9, to send a KILL signal instead of a TERM signal
System information	df	Display disk space	df –H, to show the number of occupied blocks in human-readable format
	free	Display information about RAM and swap space usage	free –k, to display information about RAM and swap space usage in kilobytes

followed by the name of the command, for example, **man ls**, which will show how to list files in various ways.

1.2.3 File Access Permission

On Linux and other Unix-like operating systems, there is a set of rules for each file, which defines who can access that file and how they can access it. These rules are called *file permissions* or *file modes*. The command name chmod stands for *change mode*, and it is used to define the way a file can be accessed. For example, if one issues a command line to a file named Mary.py like chmod 765 Mary.py, the permission is indicated by -rwxrw-r-x, which allows the user to read (r), write (w), and execute (x), the group to read and write, and any other to read and execute the file. The chmod numerical format (octal modes) is presented in Table 1.2.

1.2.4 Linux Text Editors

Text editors are needed to write scripts. There are a number of available text editors such as Emacs, Eclipse, gEdit, Nano, Pico, and Vim. Here we briefly introduce Vim, a very popular Linux text editor. Vim is the editor of choice for many developers and power users. It is based on the vi editor written by Bill Joy in the 1970s for a version of UNIX. It inherits the key bindings of vi, but also adds a great deal of functionality and extensibility that are missing from the original vi. You can start Vim editor by typing vim followed with a file name. After you finish the text file, you can type

TABLE 1.2   The chmod Numerical Format (Octal Modes)

Number	Permission	rwx
7	Read, write, and execute	111
6	Read and write	110
5	Read and execute	101
4	Read only	100
3	Write and execute	011
2	Write only	010
1	Execute only	001
0	None	000

semicolon (:) plus a lower case letter x to save the file and exit Vim editor. Table 1.3 lists the most common basic commands used in the Vim editor.

1.2.5 Keyboard Shortcuts

The command line can be quite powerful, but typing in long commands or file paths is a tedious process. Here are some shortcuts that will have you running long, tedious, or complex commands with just a few key-strokes (Table 1.4). If you plan to spend a lot of time at the command line, these shortcuts will save you a ton of time by mastering them. One should become a computer ninja with these time-saving shortcuts.

1.2.6 Write Shell Scripts

A shell script is a computer program or series of commands written in plain text file designed to be run by the Linux/Unix shell, a command-line interpreter. Shell scripts can automate the execution of repeated tasks and save lots of time. Shell scripts are considered to be scripting languages

TABLE 1.3   Common Basic Vim Commands

Key	Description
h	Moves the cursor one character to the left
l	Moves the cursor one character to the right
j	Moves the cursor down one line
k	Moves the cursor up one line
o	Moves the cursor to the beginning of the line
$	Moves the cursor to the end of the line
w	Move forward one word
b	Move backward one word
G	Move to the end of the file
gg	Move to the beginning of the file

TABLE 1.4   Common Linux Keyboard Shortcut Commands

Key	Description
Tab	Autocomplete the command if there is only one option
↑	Scroll and edit the command history
Ctrl + d	Log out from the current terminal
Ctrl + a	Go to the beginning of the line
Ctrl + e	Go to the end of the line
Ctrl + f	Go to the next character
Ctrl + b	Go to the previous character
Ctrl + n	Go to the next line
Ctrl + p	Go to the previous line
Ctrl + k	Delete the line after cursor
Ctrl + u	Delete the line before cursor
Ctrl + y	Paste

or programming languages. The many advantages of writing shell scripts include easy program or file selection, quick start, and interactive debugging. Above all, the biggest advantage of writing a shell script is that the commands and syntax are exactly the same as those directly entered at the command line. The programmer does not have to switch to a totally different syntax, as they would if the script was written in a different language or if a compiled language was used. Typical operations performed by shell scripts include file manipulation, program execution, and printing text. Generally, three steps are required to write a shell script: (1) Use any editor like Vim or others to write a shell script. Type vim first in the shell prompt to give a file name first before entering the vim. Type your first script as shown in Figure 1.2a, save the file, and exit Vim. (2) Set execute

```
(a)
#
# My first shell script
#
clear
echo "Next generation DNA sequencing increases the speed and reduces the cost of
  DNA sequencing relative to the first generation DNA sequencing."

(b)
Next generation DNA sequencing increases the speed and reduces the cost of DNA
  sequencing relative to the first generation DNA sequencing
```

FIGURE 1.2   Example of a shell script using Vim editor (a) and print out of the script after execution (b).

permission for the script as follows: chmod 765 first, which allows the user to read (r), write (w), and execute (x), the group to read and write, and any other to read and execute the file. (3) Execute the script by typing: ./first. The full script will appear as shown in Figure 1.2b.

1.3 STEP-BY-STEP TUTORIAL ON NEXT-GENERATION SEQUENCE DATA ANALYSIS BY RUNNING BASIC LINUX COMMANDS

By running Linux commands, this tutorial demonstrates a step-by-step general procedure for next-generation sequence data analysis by first retrieving or downloading a raw sequence file from NCBI/NIH Gene Expression Omnibus (GEO, http://www.ncbi.nlm.nih.gov/geo/); second, exercising quality control of sequences; third, mapping sequencing reads to a reference genome; and fourth, visualizing data in a genome browser. This tutorial assumes that a user of a desktop or laptop computer has an Internet connection and an SSH such as PuTTY, which can be logged onto a Linux-based high-performance computer cluster with needed software or programs. All the following involved commands in this tutorial are supposed to be available in your current directory, like /home/username. It should be mentioned that this tutorial only gives you a feel on next-generation sequence data analysis by running basic Linux commands and it won't cover complete pipelines for next-generation sequence data analysis, which will be detailed in subsequent chapters.

1.3.1 Step 1: Retrieving a Sequencing File

After finishing the sequencing project of your submitted samples (patient DNAs or RNAs) in a sequencing core or company service provider, often you are given a URL or ftp address where you can download your data. Alternatively, you may get sequencing data from public repositories such as NCBI/NIH GEO and Short Read Archives (SRA, http://www.ncbi.nlm. nih.gov/sra). GEO and SRA make biological sequence data available to the research community to enhance reproducibility and allow for new discoveries by comparing data sets. The SRA store raw sequencing data and alignment information from high-throughput sequencing platforms, including Roche 454 GS System®, Illumina Genome Analyzer®, Applied Biosystems SOLiD System®, Helicos Heliscope®, Complete Genomics®, and Pacific Biosciences SMRT®. Here we use a demo to retrieve a short-read sequencing file (SRR805877) of breast cancer cell lines from the experiment series (GSE45732) in NCBI/NIH GEO site.

1.3.1.1 Locate the File

Go to the GEO site (http://www.ncbi.nlm.nih.gov/geo/) → select Search GEO Datasets from the dropdown menu of Query and Browse → type GSE45732 in the Search window → click the hyperlink (Gene expression analysis of breast cancer cell lines) of the first choice → scroll down to the bottom to locate the SRA file (SRP/SRP020/SRP020493) prepared for ftp download → click the hyperlynx(ftp) to pinpoint down the detailed ftp address of the source file (SRR805877, ftp://ftp-trace.ncbi.nlm.nih.gov/sra/sra-instant/reads/ByStudy/sra/SRP%2FSRP020%2FSRP020493/SRR805877/).

1.3.1.2 Downloading the Short-Read Sequencing File (SRR805877) from NIH GEO Site

Type the following command line in the shell prompt: "wget ftp://ftp-trace.ncbi.nlm.nih.gov/sra/sra-instant/reads/ByStudy/sra/SRP%2FSRP020%2FSRP020493 /SRR805877/SRR805877.sra."

1.3.1.3 Using the SRA Toolkit to Convert .sra Files into .fastq Files

FASTQ format is a text-based format for storing both a biological sequence (usually nucleotide sequence) and its corresponding quality scores. It has become the *de facto* standard for storing the output of high-throughput sequencing instruments such as the Illumina's HiSeq 2500 sequencing system. Type "fastq-dump SRR805877.sra" in the command line. SRR805877.fastq will be produced. If you download paired-end sequence data, the parameter "-I" appends read id after spot id as "accession.spot.readid" on defline and the parameter "--split-files" dump each read into a separate file. Files will receive a suffix corresponding to its read number. It will produce two fastq files (--split-files) containing ".1" and ".2" read suffices (-I) for paired-end data.

1.3.2 Step 2: Quality Control of Sequences

Before doing analysis, it is important to ensure that the data are of high quality. FASTQC can import data from FASTQ, BAM, and Sequence Alignment/Map (SAM) format, and it will produce a quick overview to tell you in which areas there may be problems, summary graphs, and tables to assess your data.

1.3.2.1 Make a New Directory "Fastqc"

At first, type "mkdir Fastqc" in the command line, which will build Fastqc directory. Fastqc directory will contain all Fastqc results.

1.3.2.2 Run "Fastqc"

Type "fastqc -o Fastqc/SRR805877.fastq" in the command line, which will run Fastqc to assess SRR805877.fastq quality. Type "Is -l Fastqc/," you will see the results in detail.

1.3.3 Step 3: Mapping Reads to a Reference Genome

At first, you need to prepare genome index and annotation files. Illumina has provided a set of freely downloadable packages that contain bowtie indexes and annotation files in a general transfer format (GTF) from UCSC Genome Browser Home (genome.ucsc.edu).

1.3.3.1 Downloading the Human Genome and Annotation from Illumina iGenomes

Type "wget ftp://igenome:G3nom3s4u@ussd-ftp.illumina.com/Homo_sapiens/UCSC/hg19/Homo_sapiens_UCSC_hg19.tar.gz" and download those files.

1.3.3.2 Decompressing .tar.gz Files

Type "tar -zxvf Homo_sapiens_Ensembl_GRCh37.tar.gz" for extracting the files from archive.tar.gz.

1.3.3.3 Link Human Annotation and Bowtie Index to the Current Working Directory

Type "In -s homo.sapiens/UCSC/hg19/Sequence/WholeGenomeFasta/genome.fa genome.fa"; type "In -s homo.sapiens/UCSC/hg19/Sequence/Bowtie2Index/genome.1.bt2 genome.1.bt2"; type "In -s homo.sapiens/UCSC/hg19/Sequence/Bowtie2Index/genome.2.bt2 genome.2.bt2"; type "In -s homo.sapiens/UCSC/hg19/Sequence/Bowtie2Index/genome.3.bt2 genome.3.bt2"; type "In -s homo.sapiens/UCSC/hg19/Sequence/Bowtie2Index/genome.4.bt2 genome.4.bt2"; type "In -s homo.sapiens/UCSC/hg19/Sequence/Bowtie2Index/genome.rev.1.bt2 genome.rev.1.bt2"; type "In -s homo.sapiens/UCSC/hg19/Sequence/Bowtie2Index/genome.rev.2.bt2 genome.rev.2.bt2"; and type "In -s homo.sapiens/UCSC/hg19/Annotation/Genes/genes.gtf genes.gtf."

1.3.3.4 Mapping Reads into Reference Genome

Type "mkdir tophat" in the command line to create a directory that contains all mapping results. Type "tophat -p 8 -G genes.gtf -o tophat/genome SRR805877.fastq" to align those reads to human genome.

1.3.4 Step 4: Visualizing Data in a Genome Browser

The primary output of TopHat are the aligned reads BAM file and junctions BED file, which allows read alignments to be visualized in genome browser. A BAM file (*.bam) is the compressed binary version of a SAM file that is used to represent aligned sequences. BED stands for Browser Extensible Data. A BED file format provides a flexible way to define the data lines that can be displayed in an annotation track of the UCSC Genome Browser. You can choose to build a density graph of your reads across the genome by typing the command line: "genomeCoverageBed -ibam tophat/accepted_hits.bam -bg -trackline -trackopts 'name="SRR805877" color=250,0,0'>SRR805877.bedGraph" and run. For convenience, you need to transfer these output files to your desktop computer's hard drive.

1.3.4.1 Go to Human (Homo sapiens) Genome Browser Gateway

You can load bed or bedGraph into the UCSC Genome Browser to visualize your own data. Open the link in your browser: http://genome.ucsc.edu/cgi-bin/hgGateway?hgsid=409110585_zAC8Aks9YLbq7YGhQiQtwnOhoRfX&clade=mammal&org=Human&db=hg19.

1.3.4.2 Visualize the File

Click on add custom tracks button → click on Choose File button, and select your file → click on Submit button → click on go to genome browser. BED files will provide the coordinates of regions in a genome; most basically chr, start, and end. bedGraph files can give coordinate information as in BED files and coverage depth of sequencing over a genome.

BIBLIOGRAPHY

1. Haas, J. *Linux, the Ultimate Unix*, 2004, http://linux.about.com/cs/linux101/a/linux_2.htm.
2. Gite, VG. *Linux Shell Scripting Tutorial v1.05r3-A Beginner's Handbook*, 1999–2002, http://www.freeos.com/guides/lsst/.
3. Brockmeier, J.Z. *Vim 101: A Beginner's Guide to Vim*, 2009, http://www.linux.com/learn/tutorials/228600-vim-101-a-beginners-guide-to-vim.
4. Chris Benner et al. *HOMER (v4.7), Software for motif discovery and next generation sequencing analysis*, August 25, 2014, http://homer.salk.edu/homer/basicTutorial/.
5. Shotts, WE, Jr. *The Linux Command Line: A Complete Introduction*, 1st ed., No Starch Press, January 14, 2012.
6. Online listing of free Linux books. http://freecomputerbooks.com/unix-LinuxBooks.html.

## CHAPTER 2

Python for Big Data Analysis

Dmitry N. Grigoryev

CONTENTS

2.1 INTRODUCTION TO PYTHON

Python is a powerful, flexible, open-source programming language that is easy to use and easy to learn. With the help of Python you will be able to manipulate large data sets, which is hard to do with common data operating programs such as Excel. But saying this, you do not have to give up your friendly Excel and its familiar environment! After your Big Data manipulation with Python is completed, you can convert results back to your favorite Excel format. Of course, with the development of technology at some point, Excel would accommodate huge data files with all known genetic variants, but the functionality and speed of data processing by Python would be hard to match. Therefore, the basic knowledge of programming in Python is a good investment of your time and effort. Once you familiarize yourself with Python, you will not be confused with it or intimidated by numerous applications and tools developed for Big Data analysis using Python programming language.

2.2 APPLICATION OF PYTHON

There is no secret that the most powerful Big Data analyzing tools are written in compiled languages like C or java, simply because they run faster and are more efficient in managing memory resources, which is crucial for Big Data analysis. Python is usually used as an auxiliary language and serves as a *pipeline glue*. The TopHat tool is a good example of it [1]. TopHat consists of several smaller programs written in C, where Python is employed to interpret the user-imported parameters and run small C programs in sequence. In the tutorial section, we will demonstrate how to glue together a pipeline for an analysis of a FASTQ file.

However, with fast technological advances and constant increases in computer power and memory capacity, the advantages of C and java have become less and less obvious. Python-based tools have started taking over because of their code simplicity. These tools, which are solely based on Python, have become more and more popular among researchers. Several representative programs are listed in Table 2.1.

As you can see, these tools and programs cover multiple areas of Big Data analysis, and number of similar tools keep growing.

2.3 EVOLUTION OF PYTHON

Python's role in bioinformatics and Big Data analysis continues to grow. The constant attempts to further advance the first-developed and most popular set of Python tools for biological data manipulation, Biopython (Table 2.1), speak volumes. Currently, Biopython has eight actively developing projects (http://biopython.org/wiki/Active_projects), several of which will have potential impact in the field of Big Data analysis.

TABLE 2.1   Python-Based Tools Reported in Biomedical Literature

Tool	Description	Reference
Biopython	Set of freely available tools for biological computation	Cock et al. [2]
Galaxy	An open, web-based platform for data intensive biomedical research	Goecks et al. [3]
msatcommander	Locates microsatellite (SSR, VNTR, &c) repeats within FASTA-formatted sequence or consensus files	Faircloth et al. [4]
RseQC	Comprehensively evaluates high-throughput sequence data especially RNA-seq data	Wang et al. [5]
Chimerascan	Detects chimeric transcripts in high-throughput sequencing data	Maher et al. [6]

The perfect example of such tool is a development of a generic feature format (GFF) parser. GFF files represent numerous descriptive features and annotations for sequences and are available from many sequencing and annotation centers. These files are in a TAB delimited format, which makes them compatible with Excel worksheet and, therefore, more friendly for biologists. Once developed, the GFF parser will allow analysis of GFF files by automated processes.

Another example is an expansion of Biopython's population genetics (PopGen) module. The current PopGen tool contains a set of applications and algorithms to handle population genetics data. The new extension of PopGen will support all *classic* statistical approaches in analyzing population genetics. It will also provide extensible, easy-to-use, and future-proof framework, which will lay ground for further enrichment with newly developed statistical approaches.

As we can see, Python is a living creature, which is gaining popularity and establishing itself in the field of Big Data analysis. To keep abreast with the Big Data analysis, researches should familiarize themselves with the Python programming language, at least at the basic level. The following section will help the reader to do exactly this.

2.4 STEP-BY-STEP TUTORIAL OF PYTHON SCRIPTING IN UNIX AND WINDOWS ENVIRONMENTS

Our tutorial will be based on the real data (FASTQ file) obtained with Ion Torrent sequencing (www.lifetechnologies.com). In the first part of the tutorial, we will be using the UNIX environment (some tools for processing FASTQ files are not available in Windows). The second part of the tutorial can be executed in both environments. In this part, we will revisit the pipeline approach described in the first part, which will be demonstrated in the Windows environment. The examples of Python utility in this tutorial will be simple and well explained for a researcher with biomedical background.

2.4.1 Analysis of FASTQ Files

First, let us install Python. This tutorial is based on Python 3.4.2 and should work on any version of Python 3.0 and higher. For a Windows operation system, download and install Python from https://www.python .org/downloads. For a UNIX operating system, you have to check what version of Python is installed. Type python -V in the command line, if the version is below 3.0 ask your administrator to update Python and also ask to have the reference genome and tools listed in Table 2.2 installed. Once we have everything in place, we can begin our tutorial with the introduction to the pipelining ability of Python. To answer the potential question of why we need pipelining, let us consider the following list of required commands that have to be executed to analyze a FASTQ file. We will use a recent publication, which provides a resource of benchmark SNP data sets [7] and a downloadable file bb17523_PSP4_BC20.fastq from ftp://ftp-trace.ncbi.nih.gov/giab/ftp/data/NA12878/ion_exome. To use this file in our tutorial, we will rename it to test.fastq.

In the meantime, you can download the human hg19 genome from Illumina iGenomes (ftp://igenome:G3nom3s4u@ussd-ftp.illumina.com/Homo_sapiens/UCSC/hg19/Homo_sapiens_UCSC_hg19.tar.gz). The files are zipped, so you need to unpack them.

In Table 2.2, we outline how this FASTQ file should be processed.

Performing the steps presented in Table 2.2 one after the other is a laborious and time-consuming task. Each of the tools involved will take somewhere from 1 to 3 h of computing time, depending on the power of your computer. It goes without saying that you have to check on the progress of your data analysis from time to time, to be able to start the next step. And, of course, the overnight time of possible computing will be lost, unless somebody is monitoring the process all night long. The pipelining with Python will avoid all these trouble. Once you start your pipeline, you can forget about your data until the analysis is done, and now we will show you how.

For scripting in Python, we can use any text editor. Microsoft (MS) Word will fit well to our task, especially given that we can trace the whitespaces of

TABLE 2.2 Common Steps for SNP Analysis of Next-Generation Sequencing Data

Step	Tool	Goal	Reference
1	Trimmomatic	To trim nucleotides with bad quality from the ends of a FASTQ file	Bolger et al. [8]
2	PRINSEQ	To evaluate our trimmed file and select reads with good quality	Schmieder et al. [9]
3	BWA-MEM	To map our good quality sequences to a reference genome	Li et al. [10]
4	SAMtools	To generate a BAM file and sort it	Li et al. [11]
5		To generate a MPILEUP file	
6	VarScan	To generate a VCF file	Koboldt et al. [12]

our script by making them visible using the formatting tool of MS Word. Open a new MS Word document and start programming in Python! To create a pipeline for analysis of the FASTQ file, we will use the Python collection of functions named subprocess and will import from this collection function *call*.

The first line of our code will be

```
from subprocess import call
```

Now we will write our first pipeline command. We create a variable, which you can name at will. We will call it step_1 and assign to it the desired pipeline command (the pipeline command should be put in quotation marks and parenthesis):

```
step_1 = ("java -jar ~/programs/Trimmomatic-0.32/
trimmomatic-0.32.jar SE -phred33 test.fastq test_trmd.
fastq LEADING:25 TRAILING:25 MINLEN:36")
```

Note that a single = sign in programming languages is used for an assignment statement and not as an *equal* sign. Also note that whitespaces are very important in UNIX syntax; therefore, do not leave any spaces in your file names. Name your files without spaces or replace spaces with underscores, as in test_trimmed.fastq. And finally, our Trimmomatic tool is located in the *programs* folder, yours might have a different location. Consult your administrator, where all your tools are located.

Once our first step is assigned, we would like Python to display variable step_1 to us. Given that we have multiple steps in our pipeline, we would like to know what particular step our pipeline is running at a given time. To trace the data flow, we will use print() function, which will display on the monitor what step we are about to execute, and then we will use call() function to execute this step:

```
print(step_1)
call(step_1, shell = True)
```

Inside the function call() we have to take care of the shell parameter. We will assign shell parameter to True, which will help to prevent our script from tripping over whitespaces, which might be encountered on the path to the location of your Trimmomatic program or test.fastq file. Now we will build rest of our pipeline in the similar fashion, and our final script will look like this:

```
from subprocess import call
step_1 = ("java -jar ~/programs/Trimmomatic-0.32/
trimmomatic-0.32.jar SE -phred33 test.fastq test_
trimmed.fastq LEADING:25 TRAILING:25 MINLEN:36")
print(step_1)
call(step_1, shell = True)
step_2 = ("perl ~/programs/prinseq-lite-0.20.4/
prinseq-lite.pl -fastq test_trimmed.fastq -min_qual_
mean 20 -out_good test_good")
print(step_2)
call(step_2, shell = True)
step_3 = ("bwa mem -t 20 homo.sapiens/UCSC/hg19/
Sequence/BWAIndex/genome.fa test_good.fastq > test_
good.sam")
print(step_3)
call(step_3, shell = True)
step_4 = ("samtools view -bS test_good.sam > test_
good.bam")
print(step_4)
call(step_4, shell = True)
step_5 = ("samtools sort test_good.bam
test_good_sorted")
print(step_5)
call(step_5, shell = True)
step_6 = ("samtools mpileup -f homo.sapiens/UCSC/
hg19/Sequence/WholeGenomeFasta/genome.fa test_good_
sorted.bam > test_good.mpileup")
print(step_6)
call(step_6, shell = True)
step_7 = ("java -jar ~/programs/VarScan.v2.3.6.jar
mpileup2snp test_good.mpileup --output-vcf 1 >
test.vcf")
print(step_7)
call(step_7, shell = True)
```

Now we are ready to go from MS Word to a Python file. In UNIX, we will use vi text editor and name our Python file pipeline.py, where extension .py will tell that this is a Python file.

In UNIX command line type: vi pipeline.py

The empty file will be opened. Hit i on your keyboard and you will activate the INSERT mode of the vi text editor. Now copy our whole script from MS Word into pipeline.py file. Inside the vi text editor, click right mouse button and select from the popup menu Paste. While inside the vi text editor, turn off the INSERT mode by pressing the Esc key. Then type ZZ, which will save and close pipeline.py file. The quick tutorial for the vi text editor can be found at http://www.tutorialspoint.com/unix/unix-vi-editor.htm.

Once our pipeline.py file is created, we will run it with the command:

```
python pipeline.py
```

This script is universal and should processs any FASTQ file.

2.4.2 Analysis of VCF Files

To be on the same page with those who do not have access to UNIX and were not able to generate their own VCF file, we will download the premade VCF file TSVC_variants.vcf from the same source (ftp://ftp-trace.ncbi.nih. gov/giab/ftp/data/NA12878/ion_exome), and will rename it to test.vcf.

From now on we will operate on this test.vcf file, which can be analyzed in both UNIX and Windows environments. You can look at this test.vcf files using the familiar Excel worksheet. Any Excel version should accommodate our test.vcf file; however, if you try to open a bigger file, you might encounter a problem. Excel will tell that it cannot open the whole file. If you wonder why, the answer is simple. If, for example, you are working with MS Excel 2013, the limit of rows for a worksheet in this version will be 1,048,576. It sounds like a lot, but wait, to accommodate all SNPs from the whole human genome the average size of a VCF file will need to be up to 1,400,000 rows [13]. Now you realize that you have to manipulate your file by means other than Excel. This is where Python becomes handy. With its help you can reduce the file size to manageable row numbers and at the same time retain meaningful information by excluding rows without variant calls.

First, we will remove the top rows of the VCF file, which contain description entries. In the output of our pipeline, these entries occupy 64 rows. You can examine those entries using a partially opened file in Excel. Of course, you can delete them, but after saving this file you will lose rows that did not fit the worksheet. To deal with this problem, we will create a simple script using Python. There are two ways to script for it. The first approach would be to cut out the exact number of rows (in our case 64). To do our scripting, we again will use MS Word and will start with telling Python what file we are going to operate on. We will assign our test. vcf file to a variable. The assigned variable can be named whatever your fantasy desires. We will keep it simple and name it file. Now we will use function open() to open our file. To make sure that this file will not be accidently altered in any way, we will use argument of open() function 'r', which allows Python only to read this file. At the same time, we will create an output file and call it newfile. Again, we will use function open() to create our new file with name test_no_description_1.vcf. To tell Python that it can write to this file, we will use argument of open() function 'w':

```
file = open("test.vcf",'r')
newfile = open("test_no_description_1.vcf",'w')
```

Now we will create all variables that are required for our task. In this script, we will need only two of them. One we will call line and the other— *n*, where line will contain information about components of each row in test.vcf, and *n* will contain information about the sequential number of a row. Given that line is a string variable (contains string of characters), we will assign to it any string of characters of your choosing. Here we will use "abc." This kind of variable is called *character variable* and its content should be put in quotation marks. The *n* variable on the other hand will be a *numeric variable* (contains numbers); therefore, we will assign a number to it. We will use it for counting rows, and given that we do not count any rows yet, we assign 0 to *n* without any quotation marks.

```
line = "abc"
n = 0
```

Now we are ready for the body of the script. Before we start, we have to outline the whole idea of the script function. In our case, the script should read the test.vcf file line by line and write all but the first 64 lines to a new file. To read the file line by line, we need to build a repetitive structure—in programming world this is called *loops*. There are several loop structures in Python, for our purpose we will use the "while" structure. A Python while loop behaves quite similarly to common English. Presumably, you would count the pages of your grant application. If a page is filled with the text from top to bottom, you would count this page and go to the next page. As long as your new page is filled up with the text, you would repeat your action of turning pages until you reach the empty page. Python has a similar syntax: while line != "":

This line of code says: while the content of a line is not empty (does not equal [!=] empty quotation marks) do what you are asked to do in the next block of code (body of the loop). Note that each statement in Python (in our case looping statement) should be completed with the colon sign (:). Actually, this is the only delimiter that Python has. Python does not use delimiters such as curly braces to mark where the function code starts and stops as in other programming languages. What Python uses instead is indentations. Blocks of code in Python are defined by their indentation. By *block of code*, in our case we mean the content of the body of our "while" loop. Indenting the starts of a block and unindenting ends it. This means that whitespaces in Python are significant and must be consistent. In our example, the code of loop body will be indented six spaces. It does not need to be exactly six spaces, it has to be at least one, but once you have selected your indentation size, it needs to be consistent. Now we are going to populate out while loop. As we have decided above, we have to read the content of the first row from test.vcf. For this we will use function readline(). This function should be attached to a file to be read via a point sign. Once evoked, this function reads the first line of provided file into variable line and automatically jumps to the next line in the file.

```
line = file.readline()
n = n + 1
```

To keep track of numbers for variable line, we started up our counter n. Remember, we set *n* to 0. Now our *n* is assigned number 1, which corresponds to our row number. With each looping, *n* will be increasing by 1 until the loop reaches the empty line, which is located right after the last populated line of test.vcf.

Now we have to use another Python structure: if-else statement.

```
if n <= 64:
    continue
else:
    newfile.write(line)
```

Again, this block of code is close to English. If your whole program project grant application is less than or equal to 64 pages, you are fine; otherwise (else), you have to cut the text. In our case, if number of rows is below or equal to 64, we will do nothing. This is exactly what the key word continue stands for. It tells Python to stop doing anything further and come back to the top of the "while" loop and continue to read new line from the test.vcf file. Note that the if statement is completed with the colon (:). The block of if statement (in our case continue) is indented, which means that it will be executed only when the condition in the if statement is true. Once we went over the line number 64, we want the rest of the test.vcf file to be written to our new file. Here we used the write() function. As with readline() function, we attached write() function to a file to be written to via a point sign. Inside of the parenthesis of a function, we put the argument line to let the function know what to write to the newfile. Note that the logical else statement is also completed with the colon (:). The block of else statement (newfile.write(line) in our case) is indented, which means that it will be executed only when the original condition, if $n <= 64$, is false. In an if-else statement, only one of two indented blocks can be executed. Once we run our loop and generated a file, which does not have 64 descriptive rows in it, we can close both original and newly generated files. To do this, we will use function close(). Once again, we will attach close() function to a file to be closed via a point sign.

```
newfile.close()
file.close()
```

Note that there is no indentation for these lines of code and they are aligned with the while line != "": statement. It tells Python that these two lines of code are not a part of the while loop; therefore, it will be executed in the normal forward flow of statements, after the while statement is completed. Now we will glue our scripts together. To know whether our code was executed from the start to the end, we will use function print(), which will display on the computer screen whatever we will put inside the parenthesis as an argument. We put "START" and "END" into it, which will complete our script. Now we can copy and paste our script into the vi text editor as described above. At this point, we also can start using Python graphical user interface (GUI) Shell designed for Windows. We assume that you have already installed Python on your Windows machine and created a shortcut icon on your desktop. Double click the icon. If you do not have a shortcut icon, start Python by selecting START → Programs → Python 3.4 → IDLE (Python 3.4 GUI). Python shell will be opened. Inside the shell, select File → New file or hit Ctrl + N. The new Untitled window will appear. This is where we are going to paste our script (Edit → Paste or Ctrl + V) and save it (File → Save or Ctrl + S) as step_1a.py (Figure 2.1). To run this script, select Run → Run Module or hit F5.

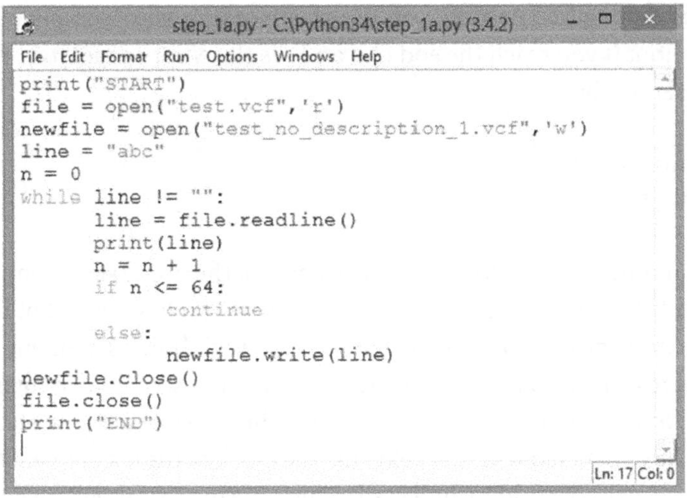

```
step_1a.py - C:\Python34\step_1a.py (3.4.2)        -  □   ×

File  Edit  Format  Run  Options  Windows  Help
print("START")
file = open("test.vcf",'r')
newfile = open("test_no_description_1.vcf",'w')
line = "abc"
n = 0
while line != "":
        line = file.readline()
        print(line)
        n = n + 1
        if n <= 64:
                continue
        else:
                newfile.write(line)
newfile.close()
file.close()
print("END")

                                                    Ln: 17 Col: 0
```

FIGURE 2.1   Python GUI (graphical user interface) with step_1a.py script.

Make sure that your step_1a.py file and test.vcf file are located in the same directory. Once we have familiarized ourselves with Python scripting, we will move to a more complex task. As we said above, there are two ways to code for removing descriptive rows from a VCF file. One can ask: why do we need another approach to perform this file modification? The answer is: not all VCF files are created in the same way. Although, by convention, all descriptive rows in VCF files begin with double pound sign (##), the number of descriptive rows varies from one sequence aligning program to another. For instance, VCF files generated by Genome Analysis Toolkit for FASTQ files from Illumina platform have 53 descriptive rows [13] and our pipeline described above will generate 23 descriptive rows. Of course, we can change our logical statement if $n <= 64$: to if $n <= 53$: or if $n <= 23$:, but why do not make our code universal? We already know that each descriptive row in VCF files begin with ## sign; therefore, we can identify and remove them. Given that we are planning to manipulate on the row content, we have to modify our loop. Our previous script was taking the empty row at the end of the test.vcf file and was writing it to the test_no_description_1.vcf file without looking into the row content. Now, when we operate on the content of a row, Python will complain about the empty content and will report an error. To avoid this, we have to make sure that our script does not operate with the empty row. To do this, we will check whether the row is empty beforehand, and if it is, we will use break statement to abort our script. Once again, our code will be close to English. Assume you are proofreading your completed grant application. If you reach the end of it and see the empty page, you are done and deserve a break.

```
if line == "":
    break
```

As you might have noticed, we used just part of the if-else statement, which is perfectly legal in Python. Once our program reaches the end of the file, there is nothing else to do but stop the script with break statement; therefore, there is no need for any else. And another new sign double equal (==) stands for a regular *equal*. Note that even the shortened if-else statement should be completed with the colon (:). The block of the if statement (in our case break) also should be indented, which means that it will be executed only when the condition in the if statement is true. Now, when we created internal break, we do not need the redundant check point at the beginning of our loop. Therefore, we will replace while line != "": with while 1:. Here we have introduced the "infinite" loop. The statement while 1 will run our loop forever unless we stop it with a break statement. Next, we will modify our existing if-else statement. Given that now we are searching for a ## pattern inside the row, rather than simply counting rows, we will replace

```
if n <= 64:
with
if line[1] == "#":
```

With line[1], we introduce the process of counting row content in Python. The line variable here represents the whole row of test.vcf file. To visualize content of a line variable, you can simply display it on your computer screen with print() function using line as an argument.

```
file = open("test.vcf", 'r')
line = file.readline()
print(line)
```

The result will be the first line of the file test.vcf: ##fileformat=VCFv4.1.

Now you can count every single element of line starting with 0. The first character (#) will be assigned 0, the second character (#) will be assigned 1, the character (f) will be assigned 2, and so on. Using this content counting, we can display any element of the line variable by putting its consecutive number into square brackets, for example, command print(line[1]) will display the second # sign. This is our mark for the descriptive rows; therefore, whenever our script sees line[1] as #, it will skip this row and go to the next one. Our complete modified script will look like this now:

```
print("START")
file = open("test.vcf",'r')
newfile = open("test_no_description_2.vcf",'w')
line = "abc"
while 1:
        line = file.readline()
        if line == "":
                break
        if line[1] == "#":
                continue
        else:
                newfile.write(line)
newfile.close()
file.close()
print("END")
```

Now we can copy and paste our script either into vi text editor (UNIX) or into Python GUI Shell (Windows), save it as step_1b.py and run. Once we are done with cutting out the descriptive rows, we can further simplify and reduce the size of our VCF file. We will use our previous script as template, and for the beginning, we will change our input and output files. Given that our goal is to make original VCF file the Excel compatible, we will use text format for our output file and will add to the file name an extension .txt.

```
file = open("test_no_description_2.vcf",'r')
newfile = open("alleles.txt",'w')
```

Our alleles.txt will have not only a different extension but also a different content. The most efficient way for researchers to operate on the allelic distribution data is to know the allelic variant location and variant itself. Therefore, our new file will have only three columns: chromosome number, position of a variant on this chromosome, and allelic distribution of the variant. As a reminder of which part of our script was doing what, we will use comments inside of our script. Typically, the pound sign (#) is used in front of comment. In the programming world, it is called *commenting out* everything that follows the pound sign.

Creating columns titles for newfile

```
line = file.readline()
newfile.write("Chromosome" + "\t" + "Position" + "\t" +
"Alleles" + "\n")
```

Given that columns in text formatted files are separated by the TAB sign (\t), we separated our titles with "\t," and, as we learned by now, all textual entries in Python should have quotation marks. We also have to end the title row with the NEWLINE sign (\n), which tells Python that this row is completed and any further input should go to the next row. Once we are done with formatting our output file, we will restructure our existing if-else statement by adding new variable. This variable (we will call it rec) will keep record of each column content after we split a row by columns. To manipulate the row content on column-by-column bases, we will need a package of functions specifically designed to do exactly this. The package is called string. In our first pipeline exercise, we already had an experience with importing call function; here in the same fashion, we will import string using an identical Python statement: import string.

Now we are set to operate on the row content. Before we begin, we have to check how many columns the file we are going to operate on has. By convention, a VCF file has nine mandatory columns, and then starting with the tenth column, it will have one sample per column. For simplicity, in our tutorial, we have a VCF file with just one sample. We also have to know what kind of column separator is used in our data. By convention, a VCF file uses tab (\t) as a column separator. Armed with this knowledge, we can start scripting. First, we read the whole row from our file, assign it to the variable line, and will make sure that this line is not empty. Then, we will split the line in pieces according to columns using the string splitting function str.split():

```
line = file.readline()
if line == "":
    break
rec = str.split(line,"\t")
```

This function takes two arguments: what to split and by what delimiter to split. As you can see, we are splitting variable lines using tab (\t) as a delimiter. Now our row is split into 10 smaller rows of data. Any of them we can call by its sequential number, but remember in programming world every counting starts with 0. Therefore, the tenth column in our row for Python will be column number nine. To display the content of the column 10, where our sample sits, we will use print() function: print(rec[9]).

In our exercise, you will see on your computer screen the following row of data:

```
1/1:87:92:91:0:0:91:91:43:48:0:0:43:48:0:0
```

Before going further, we have to familiarize ourselves with the format of VCF file. For the detailed explanation of its format, the reader can use Genome 1000 website (http://samtools.github.io/hts-specs/VCFv4.2.pdf). For our purpose, we will consider only the genotype part of the VCF file, which is exactly what we are seeing on our screens right now.

Genotype of a sample is encoded as allele values separated by slash (/). The allele values are 0 for a reference allele (which is provided in the REF column—column four or rec[3]) and 1 for the altered allele (which is provided in the ALT column—column five or rec[4]). For homozygote calls, examples could be either 0/0 or 1/1, and for heterozygotes either 0/1 or 1/0. If a call cannot be made for a sample at a given locus, each missing allele will be specified with a point sign (./.). With this knowledge in hands, the reader can deduce that in order to identify genotype of a sample we are going to operate on the first and the third elements of rec[9] (row of data above), which are representing codes for alleles identified by sequencing (in our example 1 and 1 or ALT and ALT alleles, respectively). But once again for Python these are not positions 1 and 3, but rather 0 and 2; therefore, we tell Python that we would like to work on rec[9][0] and rec[9][2]. Now we are set with the values to work with and can resume scripting. First, we will get rid of all meaningless allele calls, which are coded with points instead of numbers (./.). Using the similar construction, which we used above for skipping descriptive rows, we will get this statement:

```
if rec [9] [0] == "." or rec [9] [2] == ".":
    continue
```

In plain English it says: if the first allele or the second allele is not detected by a sequencer, we are not interested in this row of data and will continue with the next row of data. Script will stop performing downstream commands; therefore, this row will not be written to our output alleles.txt file.

The script will return to the beginning of while loop and will start analysis of the next row in test_no_description_2.vcf file. Now we have to consider a situation when both alleles were identified by a sequencer and were assigned corresponding code, either 0 or 1. In this case, the script should write a new row to our output file. Therefore, we have to start to build this row beginning with the variant location in the genome. In our input file, this information is kept in the first two columns "Chromosome" and "Position," which are for Python rec[0] and rec[1].

```
newfile.write(rec[0] + "\t" + rec[1] + "\t")
```

Here we follow the same rule as for a title row and separating future columns by TAB (\t) character. Now we have to populate the third column of our output file with allelic information. Analyzing structure of our VCF file, we already figured out that reference allele is corresponding to rec[3] and altered allele is corresponding to rec[4]. Therefore, our script for writing first allele will be

```
# Working with the first allele
if rec[9][0] == "0":
    newfile.write(rec[3])
else:
    newfile.write(rec[4])
```

We comment to ourselves that we are working with the first allele (rec[9][0]). These lines of script tell that if an allele is coded by 0, it will be presented in the "Alleles" column as the reference allele (rec[3]), otherwise (else) it will be presented as the altered allele (rec[4]). And how do we know that we have only two choices? Because we have already got rid of non-called alleles (./.) and the rec[9][0] can only be 0 or 1. The second allele (rec[9][2]) will be processed in the same fashion (the only difference will be an addition of the NEWLINE character /n), and our complete script will be as follows:

```
print("START")
import string
file = open("test_no_description_2.vcf",'r')
newfile = open("alleles.txt",'w')
line = "abc"
# Creating columns titles for variable newfile
line = file.readline()
```

```
newfile.write("Chromosome" + "\t" + "Position" + "\t" +
"Alleles" + "\n")
while 1:
        line = file.readline()
        if line == "":
                break
        rec = str.split(line,"\t")
        if rec[9][0] == "." or rec[9][2] == ".":
                continue
        newfile.write(rec[0] + "\t" + rec[1] + "\t")
        # Working with the first allele
        if rec[9][0] == "0":
                newfile.write(rec[3])
        else:
                newfile.write(rec[4])
        # Working with the second allele
        if rec[9][2] == "0":
                newfile.write(rec[3] + "\n")
        else:
                newfile.write(rec[4] + "\n")
newfile.close()
file.close()
print("END")
```

Now we can copy and paste our script either into vi text editor (UNIX) or into Python GUI Shell (Windows), save it as step_2.py and run. When our VCF file is shortened by descriptive rows and rows that had no allelic calls, we will most likely be able to fit it into the Excel worksheet and manipulate with it in our familiar environment. Once you write your Python scripts for handling VCF files, you can keep reusing them by just substituting input and output file names. To make this process even less laborious, we can join our two-step approach into one script. There are two ways to handle it. The first one will be to pipeline our scripts as we described in the beginning of our tutorial.

Let's create new script both_steps_1.py. We will pipeline our scripts step1b.py and step2.py using the approach described in our first exercise.

```
from subprocess import call
command_1 = ("python step_1b.py")
print(command_1)
call(command_1, shell = True)
```

```
command_2 = ("python step_2.py")
print(command_2)
call(command_2, shell = True)
```

This script can be run in both UNIX and Windows environments. The second way to join step_1b.py and step_2.py scripts is simply to put them into one common Python script and name it both_steps_2.py. In this way, we will save some computing time and disk space, because there will be no need for generating intermediate file test_no_description_2.vcf. However, the joined script will be more complex in terms of indentation rule. We have to make sure that flow of our scripts follows the intended route. To do this, we will put allele selection block of step_2.py script under the else statement of step_1b.py script:

```
print("START")
import string
file = open("test.vcf",'r')
newfile = open("alleles_joined_script.txt",'w')
line = "abc"
#Creating columns titles for newfile
line = file.readline()
newfile.write("Chromosome" + "\t" + "Position" + "\t" +
"Alleles" + "\n")
while 1:
        line = file.readline()
        if line == "":
                break
        if line[1] == "#":
                continue
        else:
                rec = str.split(line,"\t")
                if rec[9][0] == "." or rec[9][2] == ".":
                        continue
        newfile.write(rec[0] + "\t" + rec[1] + "\t")
        #working with the first allele
        if rec[9][0] == "0":
                newfile.write(rec[3])
        else:
                newfile.write(rec[4])
        #working with the second allele
        if rec[9][2] == "0":
                newfile.write(rec[3] + "\n")
```

```
        else:
                newfile.write(rec[4] + "\n")
newfile.close()
file.close()
print("END")
```

This script can be run in both UNIX and Windows environments.

I hope you have enjoyed our tutorial and got a flavor of Python programming. You can continue educating yourselves with general (not Big Data related) tutorials of the usage of Python, which are available online. A good place to start for real examples is to read about Biopython (Table 2.1). You will find tutorials there, which use a number of real-life examples. You can come up with small projects for yourself like writing a script that analyzes GC content of a FASTA file or a script that parses a BLAST output file and filter on various criteria.

REFERENCES

1. Trapnell C, Roberts A, Goff L, Pertea G, Kim D, Kelley DR, Pimentel H, Salzberg SL, Rinn JL, Pachter L: Differential gene and transcript expression analysis of RNA-seq experiments with TopHat and Cufflinks. *Nat Protoc* 2012, **7**(3):562–578.
2. Cock PJ, Antao T, Chang JT, Chapman BA, Cox CJ, Dalke A, Friedberg I, Hamelryck T, Kauff F, Wilczynski B et al.: Biopython: Freely available Python tools for computational molecular biology and bioinformatics. *Bioinformatics* 2009, **25**(11):1422–1423.
3. Goecks J, Nekrutenko A, Taylor J: Galaxy: A comprehensive approach for supporting accessible, reproducible, and transparent computational research in the life sciences. *Genome Biol* 2010, **11**(8):R86.
4. Faircloth BC: Msatcommander: Detection of microsatellite repeat arrays and automated, locus-specific primer design. *Mol Ecol Resour* 2008, **8**(1): 92–94.
5. Wang L, Wang S, Li W: RSeQC: Quality control of RNA-seq experiments. *Bioinformatics* 2012, **28**(16):2184–2185.
6. Maher CA, Kumar-Sinha C, Cao X, Kalyana-Sundaram S, Han B, Jing X, Sam L, Barrette T, Palanisamy N, Chinnaiyan AM: Transcriptome sequencing to detect gene fusions in cancer. *Nature* 2009, **458**(7234):97–101.
7. Zook JM, Chapman B, Wang J, Mittelman D, Hofmann O, Hide W, Salit M: Integrating human sequence data sets provides a resource of benchmark SNP and indel genotype calls. *Nat Biotechnol* 2014, **32**(3):246–251.
8. Bolger AM, Lohse M, Usadel B: Trimmomatic: A flexible trimmer for Illumina sequence data. *Bioinformatics* 2014, **30**(15):2114–2120.
9. Schmieder R, Edwards R: Quality control and preprocessing of metagenomic datasets. *Bioinformatics* 2011, **27**(6):863–864.

10. Li H, Durbin R: Fast and accurate short read alignment with Burrows-Wheeler transform. *Bioinformatics* 2009, **25**(14):1754–1760.
11. Li H, Handsaker B, Wysoker A, Fennell T, Ruan J, Homer N, Marth G, Abecasis G, Durbin R: The sequence alignment/map format and SAMtools. *Bioinformatics* 2009, **25**(16):2078–2079.
12. Koboldt DC, Zhang Q, Larson DE, Shen D, McLellan MD, Lin L, Miller CA, Mardis ER, Ding L, Wilson RK: VarScan 2: Somatic mutation and copy number alteration discovery in cancer by exome sequencing. *Genome Res* 2012, **22**(3):568–576.
13. Shortt K, Chaudhary S, Grigoryev D, Heruth DP, Venkitachalam L, Zhang LQ, Ye SQ: Identification of novel single nucleotide polymorphisms associated with acute respiratory distress syndrome by exome-seq. *PLoS One* 2014, **9**(11):e111953.

## CHAPTER 3

R for Big Data Analysis

Stephen D. Simon

CONTENTS

3.1 INTRODUCTION

R is both a programming language and an environment for data analysis that has powerful tools for statistical computing and robust set of functions that can produce a broad range of publication quality graphs and figures. R is open source and easily extensible with a massive number of user-contributed packages available for download at the Comprehensive R Archive Network (CRAN).

R has its roots in a package developed by Richard Becker and John Chambers at Bell Labs in the 1970s through the 1990s known as S. The S language had several features that were revolutionary for the time: the storage of data in self-defining objects and the use of methods for those objects (Chambers 1999). A commercial version of S, S-plus, was introduced by Statistical Sciences Corporation in the 1990s and became very popular. Around the same time, Ross Ihaka and Robert Gentleman developed R, an open source version of S based on the GNU license. Because R was written mostly in C with a few FORTRAN libraries, it was easily ported to various Unix systems, the Macintosh, and eventually Microsoft Windows.

R grew rapidly in popularity and was even highlighted in a major *New York Times* article about data analytics (Vance 2009). While there is considerable debate about the relative popularity of R versus other statistical packages and programming languages, there is sufficient empirical data to show that R is currently one of the leaders in the field. For example, R is listed as the fifth most common programming language in job postings behind Java, statistical analysis system (SAS), Python, and C/C++/C# (Muenchen 2015). At the Kaggle website, it is listed as the favorite tool by more competitors than any other and is cited more than twice as often as the next closest favorite tool, MATLAB® (Kaggle 2015).

R is currently maintained by the R Foundation for Statistical Computing (www.r-project.org/foundation). It has a robust support network with hundreds of R bloggers and an annual international conference (UseR).

3.2 R APPLICATIONS

R has many features that make it ideal for modern data analysis. It has flexible storage choices, object-oriented features, easy extensibility, a powerful integrated development environment, strong graphics, and high-performance computing enhancements.

3.2.1 Flexible Storage Choices

R has all the commonly used data types (e.g., numeric, character, date, and logical). Data are assigned using "<-" though more recent versions of R allow you also to assign using "=". So x<-12 assigns the numeric value of 12 to *x*, and y=TRUE assigns the logical value of TRUE to *y*.

More than one value of the same type can be combined into a vector using the *c* (short for combine) function. So c(1,2,3) would be a numeric vector and c("a","b","c") would be a character vector. Sequential vectors can be produced using the operator or the seq function. 1:50 would produce all the numbers between 1 and 50 and seq(2,10,by=2) would produce all the even numbers up to 10.

Vectors of the same length and same type can also be combined in a matrix. Vectors of the same length (but possibly of different types) can be combined into a data frame, which is the most common format used for data analysis in R. A data frame is essentially the same as a table in database.

But the power of R comes largely from lists. A list liberates the data set from a restrictive rectangular grid. A list is an ordered set of elements that can include scalars, vectors of different types and lengths, whole data frames or matrices, or even other lists. Consider, for example, a microarray experiment. This experiment would contain genotypic information: expression levels for thousands of genes. It would also contain phenotypic information, such as demographic information of the patients themselves or information about the treatments that these patients are receiving. A third set of information might include parameters under which the microarray experiment was run. A list that contains a separate data frame for genotypic and phenotypic information and various scalars to document the experimental conditions provides a simpler and more manageable way to store these data than any flat rectangular grid.

Subsetting is a common need for most data analyses, and you can get subsets of a vector, matrix, or data frame using square brackets. So x[2] would be the second element of the vector *x* and y[1,1] would be the upper left entry (first row, first column) in the matrix or data frame *y*. If you want everything except some entries that you would exclude, place a negative sign in front. So z[-1,] would produce everything in the matrix *z* except for the very first row.

Subsets for a list are selected the same way except you use a double bracket. So u[[3]] would be the third element in the list *u*. If a list has names associated with each element, then the $ operator would select the element with that name. So u$min produces the element of the list *u* that has the name *min*.

3.2.2 Objects and Methods

Statistical analysis in R is conducted using function calls. The lm function, for example, produces a linear regression analysis. An important feature of R, however, is that the function call does not produce *output* in the sense that a program like SAS or Statistical Package for the Social Sciences (SPSS) would. The function call creates an object of type "lm." This object is a list with a pre-specified structure that includes a vector of regression coefficients, a vector of residuals, the QR decomposition of the matrix of independent variables, the original function call, and other information relevant to the regression analysis.

Just about every object in R has multiple methods associated with it. Most objects that store information from a statistical analysis will have print, plot, and summary methods associated with them. The plot function for lm objects, for example, will display a graph of the fitted values versus the residuals, a normal probability plot of the residuals, a scale-location plot to assess heteroscedasticity, and a plot of leverage versus residuals. The summary function will produce quartiles of the residuals, *t*-tests for individual coefficients, values for multiple R-squared and adjusted R-squared, and an overall F statistic. Objects in R will often utilize inheritance. The "nlm" object, which stores information from a non-linear regression model, and the "glm" object, which stores information from a generalized linear model, both inherit from the "lm" object. In other words, they store much of the same information as an "lm" object and rely on many of the same methods, but include more information and have additional methods specific to those more specialized analyses.

One of the great values in R is that when you store the information from a statistical procedure in an object, you can easily manipulate that object to extract specific features and then send this information to a different function. You can take the regression coefficients from an "lm" object, for example, and draw a trend line on your scatterplot using the abline function.

3.2.3 Extensibility

R is easily extensible and has thousands of user-contributed packages, available at the CRAN (http://cran.r-project.org/). The R language appears to be the mode by which most new methodological research in Statistics is disseminated. Of particular interest to genetics researchers is Bioconductor (http://www.bioconductor.org/), a set of packages devoted to analysis of genetic data.

There are literally thousands of R packages and you may feel that you are looking for a needle in a haystack. You should start by reading the various task views at CRAN. These task views provide brief explanations of all the R packages in a particular area like medical imaging.

While anyone can develop a user-contributed package, there are some requirements for documentation standards and software testing. The quality of these packages can still be uneven, but you should be able to trust packages that are documented in peer-reviewed publications. You should also consider the reputation of the programming team that produced the R package. Finally, the crantastic website (http://crantastic.org/) has user-submitted reviews for many R packages.

For those who want or need to use other languages and packages as part of their data analysis, R provides the packages that allow interfaces with programming languages like C++ (Rcpp) and Python (rPython); Bayesian Markov Chain Monte Carlo packages like WinBUGS (R2WinBUGS), jags (rjags, runjags), and Stan (RStan); and data mining packages like Weka (rWeka).

3.2.4 Graphics

R produces graphics that have intelligent default options. The axes, for example, are extended 4% on either end before plotting so that points at the extreme are not partially clipped. The axis labels, by default, use *pretty* values that are nice round numbers intelligently scaled to the range of the data. The default colors in R are reasonable, and you can select a wide range of color gradients for heat maps and geospatial plots using the RColorBrewer package. The par function in R allows you to adjust the graphical parameters down to the level of the length of your axis tick marks.

R is a leader in the use of data visualization based on the Grammar of Graphics Language (Wilkinson 2014), a system for creating graphs that separates graphs into semantic components. This ggplot2 package allows you to create graphs using these components, which offers you the ability to customize graphs at a high level of abstraction by changing individual components (Wickham 2010).

The CRAN graphics task view (http://cran.r-project.org/web/views/Graphics.html) shows many other R graphics packages, including a link to the lattice graphics system, which is an R implementation of the trellis system (Becker et al. 1996).

3.2.5 Integrated Development Environment

You can program in R with a basic text editor (I have used notepad with R for more years than I care to admit), but there is a very nice integrated development environment, RStudio, that you should give serious consideration to. It offers code completion, syntax highlighting, and an object browser. It integrates nicely with version control software and the R Markdown language.

R Markdown is worth a special mention. An active area of interest in the Statistics community is the concept of reproducible research. Reproducible research makes not just the data associated with a research publication available, but also the code so that other people who want to do work in this field can easily replicate all of the statistical analyses and reproduce all of the tables and graphs included in the article (Peng 2009). The R Markdown language combined with the R package knitr allows you to produce self-documenting computer output, which greatly enhances the reproducibility of your publication.

3.2.6 Size and Speed Issues

R can handle many Big Data problems without a fuss, but it has two well-known limitations. The first limitation is that loops in R are often very inefficient. The inefficiency is often not noticeable for small data sets, but loops are commonly a serious bottleneck for Big Data. You can often improve the speed of your R code by replacing the loop with a function that works equivalently (Ligges and Fox 2008). There are basic functions in R for summation (`rowSums`, `colSums`) and vector/matrix operators like `crossprod` and `outer` that will run a lot faster and improve the readability of your code. R also has a series of *apply* functions that can avoid an explicit loop. The `sapply` function, for example, takes a function that you specify and applies it to each element in a list.

If you have multiple processors available, you can sometimes split your computation into pieces that can be run in parallel. The foreach package allows you to set up a loop of independent calculations where different iterations in the loop can be run on different processors. The R code amenable to parallelization in R is limited, but this is an active area of work in the R community.

A second limitation of R is that it needs to store all available data in computer memory. These days computer memory is not trivially small, but it still has limits compared to what you can store on your local hard drive or your network. If your data are too big to fit in memory, you have to use special tools. Sometimes you can sidestep the problems with data too big to fit in memory by handling some of the data management through SQL. Other times you can reduce the size of your data set through sampling. You can also use specialized libraries that replace R functions with equivalent functions that can work with data larger than computer memory. The biglm package, for example, allows you to fit a linear regression model or a generalized linear regression model to data sets too big to fit in R.

The tools available for improving speed and storage capacity are too numerous to document here. A brief summary of the dozens of R packages that can help you is listed in the CRAN Task View on High Performance Computing (http://cran.r-project.org/web/views/HighPerformanceComputing.html).

3.2.7 Resources for Learning More About R

Since R is freely available, many of the resources for learning R are also free. You should start with the CRAN. CRAN has the official R manuals. It also has a very helpful FAQ for the overall package and a specialized FAQ for Windows, Macintosh, and Unix implementations of R. CRAN also is the repository for most R packages (excluding the packages associated with Bioconductor), and you can browse the documentation associated with each package, which is stored in a standardized format in a PDF file. Some packages have vignettes that show some worked examples.

R has a built-in help system, and selecting help from the menu will provide an overview of all the resources within R. If you want help with a particular function, type a question mark followed by the function name or use the help function. So ?plot or help("plot") provides information about the plot function. If you are not sure what the name of the function is, you can run a general search using two question marks followed by the search term, or equivalently, you can use the help.search function. So ??logistic or help.search("logistic") will help you find the function that performs logistic regression.

The best way to learn R is to practice with simple data sets. R provides a wide range of data sets with the base system and almost every package includes one or more data sets that can illustrate how the package is used. Additional data sets and code in R to analyze these data sets are available at the Institute for Digital Research and Education at the University of California at Los Angles (http://www.ats.ucla.edu/stat/dae/). This site is ideal for those already familiar with another statistics package like SAS or SPSS because you can compare the R code with the code from these other packages.

There are hundreds of R bloggers and many of them will repost their blog entries at the R-bloggers site (http://www.r-bloggers.com/). The R help mailing list is available at CRAN (http://www.r-project.org/mail.html). It is a very active list with dozens of messages per day. You may find the nabble interface to be more convenient (http://r.789695.n4.nabble.com/). Many communities have local R user groups. Revolution Analytics offers a fairly current and comprehensive list of these (http://blog.revolutionanalytics.com/local-r-groups.html).

3.3 DATA ANALYSIS OUTLINE

Every data analysis is different, but there are some common features for most analyses. First, you need to import your data. Often, you will need to manipulate your data in some way prior to data analysis. Then, you need to screen your data with some simple summary statistics and plots. For the actual data analysis, it is tempting to start with the largest and most complex model, but you should fit simpler models first, even overly simplistic models, so that you aren't immediately overwhelmed.

3.3.1 Import Your Data

R can easily handle a variety of delimited files with the `read.table` function and you can also use `read.csv`, which specializes in reading the commonly used comma separated value format, and `read.delim`, which specializes in reading tab delimited files. The foreign package allows you to import from a variety of other statistical packages (EpiInfo, SPSS, SAS, and Systat). The dbi package will connect you with most common SQL databases and there are specialized packages for Oracle (ROracle) and SQL Server (RSQLServer). You can sometimes find specialized packages to handle specific formats like MAGE-ML (RMAGEML). There are also libraries to import from various social media like Twitter (TwitteR).

3.3.2 Manipulate Your Data

R offers some very useful tools for data manipulation. The `merge` function allows you to join two separate data frames using a one-to-one or a many-to-one merge and using either an inner join or an outer join. Longitudinal data often require you to convert from a format with one record per subject and data at multiple time points strung out across horizontally to a format with multiple records per subject and one line per time point within each subject. The reshape2 package makes these conversions easy.

If you need to aggregate your data, you can choose among several different functions (apply, sapply, or tapply) depending on whether your data are stored in a matrix, a list, or a data frame. These functions are very powerful, but also rather tricky to work with. For more advanced aggregation tasks, look at the functions available in the plyr package.

Another common data manipulation is subset selection. There are several approaches, but often a simple logical expression inserted as an index within a matrix or data frame will work nicely. The grep function, which finds matches to strings or regular expressions, is another common approach for subset selection.

Many data sets have special numeric codes for categorical data. You will often find that the formal analyses will be easier to follow if you designate categorical variables using the factor function. This function also allows you to specify a label for each category value and will simplify certain regression models by treating factors as categorical rather than continuous.

3.3.3 Screen Your Data

R has a variety of tools for screening your data. The head function shows the first few lines of your matrix or data frame, and the tail function shows the last few lines. The dim function will tell you how many rows or columns your matrix/data frame has.

If your data are stored in a data frame, the summary function is very useful. It provides a list of the six most common values for string variables and factors. For variables that are numeric, summary produces the minimum, 25th percentile, median, mean, 75th percentile, and the maximum plus a count of the number of missing values if there are any. You need to watch missing values very carefully throughout the rest of your data analysis.

If you know that some of your data have only a limited number of possible values, then you should use the table function to list those values and their frequency counts. Look out for inconsistent coding, especially strings. You may find a yes/no variable that has YES, Yes, and yes for possible values and even if the person who entered the data intended for them to all mean the same thing, R will treat them as separate values.

3.3.4 Plot Your Data

R has a wide range of plotting methods. For an initial screen, you can examine simple bivariate relationships using the `plot` function. Often, a smooth curve using the `lowess` function can help you discern patterns in the plot. The `boxplot` function helps you to examine the relationship between a categorical variable and a continuous variable. For very large data sets, some data reduction technique like principal components or some data summary technique like cluster analysis may prove useful.

3.3.5 Analyze Your Data

The data analysis models are virtually unlimited, and it is impossible to summarize them here. As a general rule, you should consider the very simplest models first and then add layers of complexity slowly until you build up to your final model(s). Do this even if the simplest models are such oversimplifications that they strain the credulity of the analyses. You have to spend some time getting comfortable with the data and familiar with the general patterns that exist. The simplest models help you gain this comfort and familiarity.

3.4 STEP-BY-STEP TUTORIAL

A gene expression data set freely available on the web gives you the opportunity to try out some of the features of R. This data set, described in Son et al. (2005), includes gene expression levels for over 40,000 genes across 158 samples with 19 different organs selected from 30 different patients (not every patient contributed samples for all 19 organs). For these data, an analysis of gender differences in the gene expression levels in the adrenal gland is illustrated. The data are already normalized, reducing some of your work. The analysis suggested here is somewhat simplistic and you should consider more sophistication, both in terms of the hypothesis being addressed and in the data analysis methods. If you are unclear on how any of the functions used in this example work, review the help file on those functions.

3.4.1 Step 1: Install R, RStudio, and Any R Packages That You Need

If you do not have R already installed on your computer, go to the Comprehensive R Archive Network. Download and run the version of R appropriate for your computer.

If you wish to use the integrated development environment RStudio, download it at http://www.rstudio.com/. You will only need one R package in this example, mutoss. You can download it by running R and typing install.packages("mutoss") on the command line. You can also install mutoss from the menu system in R (select Packages|Install package(s)…).

3.4.2 Step 2: Import the Son et al. Data

The data set that you need to *import* is found in a tab delimited text file. The URL is genome.cshlp.org/content/suppl/2005/02/11/15.3.443.DC1/ Son_etal_158Normal_42k_RatioData.txt. You can read this directly from R without having to download the file, or you have the option of downloading the file to your local computer. Here is the code for reading directly.

```
file.name.part.1 <- "http://genome.cshlp.org/content/
suppl/2005/02/11/
15.3.443.DC1/"
file.name.part.2 <- "Son_etal_158Normal_42k_
RatioData.txt"
son <- read.delim(paste(file.name.part.1,file.name.
part.2,sep=""))
```

I split the filename into two parts to make it easier to modify the code if you've already downloaded the file. The paste function combines the two parts, and the read.delim function produces a data frame, which is stored in son. If you have downloaded the file, modify this code by changing from the URL address listed in file.name.part.1 to the drive and path where the downloaded file is located.

We are fortunate in this example that the file reads in easily with all the default parameters. If this did not work, you should read the help file for read.delim by typing

```
?read.delim
```

The read.delim function will produce a data frame. How big is the data frame? Use the dim command.

```
> dim(son)
[1] 42421   160
```

There are 42,421 rows and 160 columns in this data frame. Normally, I would use the head and tail functions to review the top few and bottom few rows. But with a very large number of columns, you may want to just print out the upper left and lower right corners of the data set.

```
> son[1:8,1:4]
PlatePos CloneID NS1_Adrenal NS2_Adrenal
1   CD1A1    73703         1.35         1.56
2  CD1A10   345818         1.90         4.12
3  CD1A11   418147         1.52         1.44
4  CD1A12   428103        16.81         1.48
5   CD1A2   127648        12.46        50.24
6   CD1A3    36470         0.54         0.74
7   CD1A4    37431         0.57         0.51
8   CD1A5   133762        59.26         3.98
> son[42414:42421,157:160]
      NS183_Uterus NS184_Uterus NS185_Uterus NS186_Uterus
42414         0.56         0.74         0.73         0.66
42415         0.95         0.79         1.06         0.91
42416         0.69         0.61         0.50         0.57
42417         1.73         0.92         0.57         0.98
42418         0.97         1.36         1.18         1.18
42419         0.85         1.14         0.92         0.83
42420         6.35         4.64         4.17         3.89
42421         0.64         0.57         0.56         0.52
```

3.4.3 Step 3: Select Adrenal Gland Tissue and Apply a Log Transformation

You should run several simple *data manipulations*. First, you need to select only those columns in the data associated with adrenal gland tissue. Then, you need to transform the expression levels using a base 2 logarithm. Finally, you need to identify which of the tissues belong to male and female subjects. The names command produces a character vector with the names of the variables in the data frame son. Store this in var.names. The grep function returns the column numbers which include the string *Adrenal*. son[,adrenal.columns] selects every row of son but only those columns found in adrenal.columns. Because the values of the adrenal columns are all numeric, you can convert them from a data frame to a matrix using the as.matrix function.

```
vnames <- names(son)
adrenal.columns <- grep("Adrenal",var.names)
son.a <- as.matrix(son[,adrenal.columns])
```

You should normally consider a log transformation for your data because the data are skewed and span several orders of magnitude.

```
son.a <- log(son.a,base=2)
```

There is some supplemental information stored as a PDF file that includes demographic information (gender, age, cause of death) about the patients who provided the samples. We need this file to identify which of the samples come from males and which from females. There is no easy way to directly read data from a PDF file into R. In Adobe Acrobat, select all the text, copy it to the clipboard, and paste it into a text file. This will look somewhat like a delimited file, but there are issues created when the name of the tissue type and the listing of the cause of death contain embedded blanks. This is further complicated by the lines which are blank or which contain extraneous information. So it is easier to avoid splitting each line into separate fields and instead just read in full lines of data using the readLines function. You can then select those lines that we need with the grep function, first by finding those lines containing the string *adrenal* and then searching in those lines for the string *M*. Note that the leading and trailing blanks in this string helps avoid selecting a letter M that starts, ends, or is in the middle of a word.

```
> file.name <- "Son_etal_phenotypic_information.txt"
> son.p <- readLines(file.name)
> adrenal.lines <- grep("adrenal",son.p)
> son.p <- son.p[adrenal.lines]
> males <- grepl(" M ",son.p)
> print(males)
[1] 1 2 4 5
```

We are relying here on the fact that the order listed in the PDF file is consistent with the order of the columns in the text file.

3.4.4 Step 4: Screen Your Data for Missing Values and Check the Range

There are several simple screens that you should consider. First, the is.na function will return the value TRUE for any values in the vector or matrix that are missing. If you sum that across the entire matrix, you will get a count of missing values, since TRUE converts to 1 and FALSE to 0 when you use the sum function.

```
> sum(is.na(son.a))
[1] 0
```

A zero here gives us the reassurance that the entire matrix has no missing values. The range function provides the minimum and maximum values across the entire vector or matrix.

```
> range(son.a)
[1] -8.587273 10.653669
```

This is a fairly wide range. Recall that this is a base 2 logarithm and 2 raised to the −8 power is about 0.004, while 2 raised to the 10th power is 1,024. Such a range is wide, but not unheard of for gene expression data. The summary function, when applied to a matrix or data frame, will produce percentiles and a mean for numeric data (and a count of missing values if there are any). For character data and factors, summary will list the seven most frequently occurring values along with their counts. Because of space limitations, I am showing summary only for the first two columns and the last column.

```
> summary(son.a[,c(1,2,9)])
 NS1_Adrenal          NS2_Adrenal          NS9_Adrenal
Min.    :-6.82828    Min.    :-7.96578    Min.    :-5.64386
1st Qu.:-0.66658    1st Qu.:-0.57777    1st Qu.:-0.57777
Median :-0.05889    Median : 0.01435    Median :-0.02915
Mean    :-0.02138    Mean    :-0.01424    Mean    :-0.01223
3rd Qu.: 0.54597    3rd Qu.: 0.53605    3rd Qu.: 0.47508
Max.    : 8.23903    Max.    : 8.27617    Max.    : 6.02104
```

3.4.5 Step 5: Plot Your Data

There are several plots that make sense for an initial screen of these data. You can run a simple histogram for each of the nine columns to look for unusual patterns like a bimodal distribution or expression levels that remain highly skewed even after a log transformation. Alternative patterns may still be okay, but they are a cause for further investigation. All of the histograms show a nice bell-shaped curve. Here is the histogram for the first column of data.

```
> hist(son.a[,1])
```

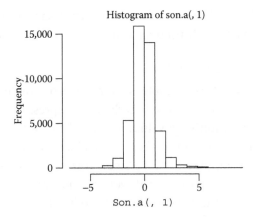

Histogram of son.a(, 1)

Another useful screen is a scatterplot. You can arrange scatterplots among all possible pairs of columns using the `pairs` function. For very large data sets, you will often find the overprinting to be a problem, and a quick fix is to change the plotting symbol from the default (a circle) to a small single pixel point.

```
> pairs(son.a,pch=".")
```

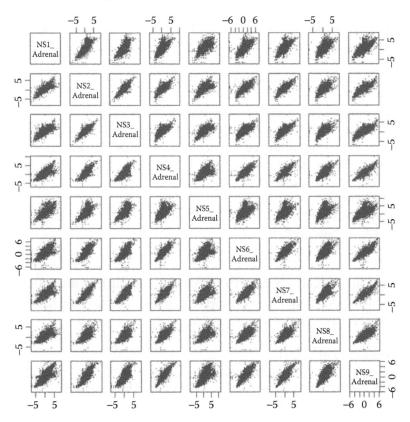

Notice that the first and fifth samples seem to depart somewhat from the overall pattern of an elliptical distribution of data, but this is not a serious concern.

3.4.6 Step 6: Analyze Your Data

The data analysis in this example is simply a two-sample *t*-test comparing males to females for each row in the gene expression matrix with an adjustment of the resulting *p*-values to control the number of tests. Let's remember the advice to wade in from the shallow end of the pool. Start by calculating a two-sample *t*-test for a single row in the data set.

If you've never run a *t*-test in R, you may not know the name of the function that does a *t*-test. Type ??ttest to list the many functions that will perform many different types of *t*-tests. The one that looks the most promising is the t.test function. Get details on this by typing ?t.test. From reading the help file, it looks like we want one group (the males) as the first argument to the t.test function and the other group (the females) as the other argument. Recall that a minus sign inverts the selection, so −males will select everyone except the males.

```
> first.row <- t.test(son.a[1,males],son.a[1,-males])
> first.row

        Welch Two Sample t-test

data:  son.a[1, males] and son.a[1, -males]
t = 0.8923, df = 9.594, p-value = 0.3941
alternative hypothesis: true difference in means is
     not equal to 0
95 percent confidence interval:
-0.1188546  0.2761207
sample estimates:
mean of x mean of y
0.527884  0.449251
```

We need to extract the *p*-value from this object for further manipulation. If you check the names of every element in this object, you will see one labeled p.value. This is what you want.

```
> names(first.row)
[1] "statistic"    "parameter"    "p.value"      "conf.int"
"estimate"     "null.value"   "alternative"  "method"
"data.name"
```

```
> first.row$p.value
[1] 0.3940679
```

Note that you could have figured this out with a careful reading of the help file on t.test. Now you need to create a special function which only extracts the *p*-value. Assign a function to t.test.pvalue using the function command. The argument(s) specified in function are arguments used in the statements contained between the two curly braces. The first statement computes a t.test using the t.test function we just tested and stores it in results. The second statement selects and returns just the *p*-value from results.

```
> t.test.pvalue <- function(dat) {
+    results <- t.test(dat[males],dat[-males])
+    return(results$p.value)
+ }
> t.test.pvalue(son.a[1,])
[1] 0.3940679
```

You can now apply this to each row of the matrix using the apply command. The first argument in apply is the matrix, the second argument specifies whether you want to extract rows (1) or columns (2), and the third argument specifies the function you wish to run on each row or column.

```
> all.rows <- apply(son.a,1,t.test.pvalue)
> head(all.rows)
[1] 0.3940679 0.5616102 0.6953087 0.3064443 0.8942156
0.8191188
> tail(all.rows)
[1] 0.8631147 0.3911861 0.4482372 0.8286146 0.8603733
0.2700229
```

Check how many of these *p*-values would be significant at a nominal alpha level of 10% with no adjustments for multiple comparisons.

```
> sum(all.rows<0.10)
[1] 1268
```

How many would be significant after a Bonferroni correction?

```
> sum(all.rows<0.10/42421)
[1] 0
```

Note that I am not recommending the use of a Bonferroni correction, not because it produced 0 significant results, but because the Bonferroni correction is considered by many to be too stringent in a gene expression study.

I ran the Bonferroni correction in spite of not liking it because it is fast and easy to understand. Remember that you need to start your analyses from the shallow end of the pool. The mutoss library has a variety of adjustments that perform better because they don't impose the excessively stringent requirement of controlling the global Type I error rate, as the Bonferroni correction does. Instead, these methods control the false discovery rate (FDR). One of the simplest methods that controls the FDR is the Benjamini–Hochberg linear step-up procedure.

First, you need to install the mutoss library. If you didn't do this already, you can type `install.packages("mutoss")`. Once this is installed, load the package with the library command.

```
> library("mutoss")
```

Now check the help file. The BH function takes a vector of *p*-values and applies the Benjamini–Hochberg adjustment procedure and controls the FDR at a specified value.

```
> bh.adjustment <- BH(pv.mf,alpha=0.1)
        Benjamini-Hochberg's (1995) step-up procedure
```

```
Number of hyp.:     42421
Number of rej.:      10
rejected        pValues       adjPValues
1   24229   1.948179e-08   0.0008264369
2   24325   1.342508e-07   0.0028475267
3   23914   4.540339e-07   0.0057640760
4   24010   5.435116e-07   0.0057640760
5   29430   1.302986e-06   0.0110547905
6   32969   2.024393e-06   0.0143127935
7   41695   5.235523e-06   0.0317280175
8   16351   8.543680e-06   0.0453039332
9    6416   1.612319e-05   0.0699810163
10 19656   1.649679e-05   0.0699810163
```

The analysis listed here is very simplistic. There's much more that you could do here. You should consider screening out some of the genes before calculating the *t*-test. You can investigate the object produced by the BH function and compare how it performs relative to other adjustments. Find the names of the 10 genes, investigate their properties, and see if these genes have a common Gene Ontology. Look for other packages that analyze gene expression data. There are, for example, packages that automate some of the steps here by combining the *t*-test with the *p*-value adjustment. Try to replicate your analysis on a different data set provided with R or with one of the R packages.

3.5 CONCLUSION

R is a powerful programming language and environment for data analysis. It has publication quality graphics and is easily extensible with a wealth of user-contributed packages for specialized data analysis.

REFERENCES

Becker RA, Cleveland WS, Shyu M (1996). The visual design and control of trellis display. *Journal of Computational and Graphical Statistics*. 5(2):123–155.

Chambers J (1999). Computing with data: Concepts and challenges. *The American Statistician*. 53(1):73–84.

Kaggle (2015). Tools used by competitors. https://www.kaggle.com/wiki/Software (Accessed March 23, 2015).

Ligges U, Fox J (2008). How can I avoid this loop or make it faster? *R News*. 8(1):46–50.

Muenchen RA (2015). The popularity of data analysis software. http://r4stats.com/articles/popularity/ (Accessed March 23, 2015).

Peng RD (2009). Reproducible research and biostatistics. *Biostatistics*. 10(3): 405–408.

Son CG, Bolke S, Davis S, Greer BT, Wei JS, Whiteford CC, Chen QR, Cenacchi N, Khan J (2005). Database of mRNA gene expression profiles of multiple human organs. *Genome Research*. 15:443–450.

Vance A (2009). Data analysts captivated by R's power. *The New York Times* (January 6).

Wickham H (2010). A layered grammar of graphics. *Journal of Computational and Graphical Statistics*. 19(1):3–28.

Wilkinson L (2005). *The Grammar of Graphics*. Springer, New York.

II

Next-Generation DNA
Sequencing Data Analysis

## CHAPTER 4

Genome-Seq Data Analysis

Min Xiong, Li Qin Zhang, and Shui Qing Ye

CONTENTS

4.1 INTRODUCTION

Genome sequencing (genome-seq, also known as whole-genome sequencing, full genome sequencing, complete genome sequencing, or entire genome sequencing) is a laboratory process that determines the complete DNA sequence of an organism's genome at a single time. This entails sequencing all of an organism's chromosomal DNA as well as DNA contained in the mitochondria and, for plants, in the chloroplast. The completion of the first human genome project has been a significant milestone in the history of medicine and biology by deciphering the order of the three billion units of DNA that go into making a human genome, as well as to identify all of the genes located in this vast amount of data. The information garnered from the human genome project has the potential to forever transform health care by fueling the hope of genome-based medicine, frequently called *personalized* or *precision medicine*, which is the future of health care. Although it is a great feat, the first $3-billion human genome project has taken more than 13 years for the completion of a reference human genome sequence using a DNA technology based on the chain terminator method or Sanger method, now considered as a first-generation DNA sequencing. Both the cost and speed in the first-generation DNA sequencing are prohibitive to sequence everyone's entire genome, a prerequisite to realize personalized or precision medicine. Since 2005, next-generation DNA sequencing technologies have been taking off, which reduce the costs of DNA sequencing by several orders of magnitude and dramatically increase the speed of sequencing. Next-generation DNA sequencing is emerging and continuously evolving as a tour de force in the genome medicine.

A number of next-generation sequencing (NGS) platforms for genome-seq and other applications have been developed. Several major NGS platforms are briefed here. Illumina's platforms (http://www.illumina.com/) represent one of the most popularly used sequencing by synthesis chemistry instruments in a massively parallel arrangement. Currently, it markets HiSeq X Five and HiSeqTen instrument with a population power; HiSeq 3000, HiSeq 4000, HiSeq 2500, and HiSeq 1500 with a production power; NextSeq 500 with a flexible power; and MiSeq with a focused power. The HiSeq X Ten is a set of 10 ultra-high-throughput sequencers, purpose-built for large-scale human whole-genome sequencing at a cost of $1000 per genome, which together can sequence over 18,000 genomes per year. The MiSeq desktop sequencer allows you to access more focused applications such as targeted gene sequencing, metagenomics, small-genome sequencing, targeted gene expression, amplicon sequencing, and HLA typing. New MiSeq reagents enable up to 15 GB of output with 25 M sequencing reads and 2 × 300 bp read lengths. Life Technologies (http://www.lifetechnologies.com/) markets sequencing by oligonucleotide ligation and detection (SOLiD) 5500 W Series Genetic Analyzers, Ion Proton™ System, and the Ion Torrent™ Personal Genome Machine® (Ion PGM™) System. The newest 5500 W instrument uses flow chips, instead of beads, to amplify templates, thus simplifying the workflow and reducing costs. Its sequencing accuracy can be up to 99.99%. Both Ion Proton™ System and Ion PGM™ System are ion semiconductor-based platforms. Ion PGM™ System is one of the top selling benchtop NGS solutions. Roche markets 454 NGS platforms (http://www.454.com/), the GS FLX+ System, and the GS Junior Plus System. They are based on sequencing by synthesis chemistry. The GS FLX+ System features the unique combination of long reads (up to 1000 bp), exceptional accuracy, and high-throughput, making the system well suited for larger genomic projects. The GS Junior Plus System is a benchtop NGS platform suitable for individual lab NGS needs. Pacific Biosciences (http://www.pacificbiosciences.com/) markets the PACBIO RSII platform. It is considered as the third-generation sequencing platform since it only requires a single molecule and reads the added nucleotides in real time. The chemistry has been termed SMRT for single molecule real time. The PacBio RS II sequencing provides average read lengths in excess of >10 KB with ultra-long reads >40 KB. The long reads are characterized by high 99.999% consensus accuracy and are ideal for de novo assembly, targeted sequencing applications, scaffolding, and spanning structural rearrangements. Oxford Nanopore Technologies (https://nanoporetech.com/) markets the GridION™ system, The PromethION, and The MinION™ devices. Nanopore sequencing is a third-generation single-molecule technique. The GridION™ system is a benchtop instrument and an electronics-based platform. This enables multiple nanopores to be measured simultaneously and data to be sensed, processed, and analyzed in real time. The PromethION is a tablet-sized benchtop instrument designed to run a small number of samples. The MinION device is a miniaturized single-molecule analysis system, designed for single use and to work through the USB port of a laptop or desktop computer. With continuous improvements and refinements, nanopore-based sequencing technology may gain its market share in the distant future.

The genetic blueprints, or genomes, of any two humans are more than 99% identical at the genetic level. However, it is important to understand the small fraction of genetic material that varies among people because it can help explain individual differences in susceptibility to disease, response to drugs, or reaction to environmental factors. Thus, knowing the full genome DNA sequence and cataloguing single-nucleotide variants and structural variants of each person are a prerequisite for personalized or precision medicine. Genome-seq has proven to be a valuable tool for detecting all genetic variants of rare and complex disease ranging from single nucleotides to larger structure. However, exome sequencing and targeted resequencing are still in use. Many monogenic and rare complex disease variants happen in coding regions, which only occupy 1%–2% of human genome. Due to the need to be more cost-effective to discover disease cause variants compared with whole-genome sequencing, exome sequencing and target resequencing are employed for a closer examination of these special regions. Exome sequencing and target resequencing both belong to targeted capture and massively parallel sequencing, and the difference is that exome-seq captures protein coding regions and target resequencing captures regions of interest, either coding regions or non-coding regions.

Sequencing goes hand in hand with computational analysis. Effective translation of the accumulating high-throughput sequence data into meaningful biomedical knowledge and application relies in its interpretation. High-throughput sequence analyses are only made possible via intelligent computational systems designed particularly to decipher meaning of the complex world of nucleotides. Most of the data obtained with state-of-the-art next-generation sequencers are in the form of short reads. Hence, analysis and interpretation of these data encounters several challenges, including those associated with base calling, sequence alignment and assembly, and variant calling. Often the data output per run are beyond the common desktop computer's capacity to handle. High power computer cluster becomes the necessity for efficient genome-seq data analysis. These challenges have led to the development of innovative computational tools and bioinformatics approaches to facilitate data analysis and clinical translation. Although de novo genome-seq is in its full swing to sequence the new genomes of animals, plants, and bacteria, this chapter only covers the human genome-seq data analysis by aligning the newly sequenced human genome data to the reference human genome. Here, we will highlight some genome-seq applications, summarize typical genome-seq data analysis procedures, and demonstrate both command-line interface-based- and graphical user interface (GUI)-based-genome-seq data analysis pipelines.

4.2 GENOME-SEQ APPLICATIONS

Whole-exome sequencing (WES) can provide coverage of more than 95% of the exons, which contains 85% of disease-causing mutations in Mendelian disorders and many disease-predisposing single-nucleotide polymorphisms (SNPs) throughout the genome. The role of more than 150 genes has been distinguished by means of WES, and this statistics is quickly growing. The decreasing cost and potential to provide a more comprehensive genetic risk assessment than current targeted methods makes whole-genome sequencing an attractive tool for genetic screening in patients with a family history of disease. Genome-seq has been picking up steams for identification of human population structure, evolution impact, causative variants, and genetic marker of Mendelian disease and complex disease, and cancer driver variations. Both WES and whole-genome sequencing (WGS) have been explored and evaluated as a fast diagnostic tool for clinical diseases. Most of applications focus on discovery of SNP, insertion or deletion of base (Indel), and structure variants (SV). Table 4.1 lists part of applications in current researches.

TABLE 4.1   Genome-Seq Applications

#	Usages	Descriptions	References
1	SNP[a]	Identifying single-nucleotide polymorphism	Genomes Project et al. (2010)
			Genomes Project et al. (2012)
2	Indel[b]	Identifying insertion or deletion of base	Genomes Project et al. (2010)
			Genomes Project et al. (2012)
3	Inversion	Identifying segment of intrachromosome reversed to end to end	Bansal et al. (2007)
4	Intrachromosomal translocation	Discovery of chromosome rearrangement in the intrachromosome	Chen et al. (2009)
5	Interchromosomal translocation	Discovery of chromosome rearrangement in the interchromosome	Chen et al. (2009)
6	CNV[c]	Identifying DNA copy number alteration	Priebe et al. (2012)
			Zack et al. (2013)
7	Gene fusion	Discovery of fusion gene	Chmielecki et al. (2013)
8	Retrotransposon	Detecting DNA elements which transcribe into RNA and reverse transcribe into DNA, and then insert into genome	Lee et al. (2012)
9	eQTL[d]	Testing association between gene expression and variation	Fairfax et al. (2014)
10	LOH[e]	Discovery of loss of the entire gene and surrounding chromosomal region	Sathirapongsasuti et al. (2011)
11	LOFS[f]	Discovery of loss of function variants	MacArthur et al. (2012)
12	Population structure and demographic inference	Using variation structure to understand migration and gene flow in population	Genome of the Netherlands et al. (2014)
13	Diagnosis of neurodevelopmental disorders	Using accelerated WGS or WES	Soden et al. (2014)

[a] SNP, single-nucleotide polymorphism.
[b] Indel, insertion or deletion of base.
[c] CNV, copy number variation.
[d] eQTL, expression quantitative trait loci.
[e] LOH, loss of heterozygosity.
[f] LOFs, loss of function variants.

4.3 OVERALL SUMMARY OF GENOME-SEQ DATA ANALYSIS

Many tools are developed for whole-genome-seq data analysis. The basic genome-seq data analysis protocol is from sequence quality control to variation calling, which shows you the number of variants and the kind of variations in your population or samples. In general, genome-seq data analysis consists of five steps as displayed in Figure 4.1 and expounded in the following texts.

> **Step 1:** *Demultiplex, filter, and trim sequencing read.* Sequencing instruments generate base call files (*.bcl) made directly from signal intensity measurements during each cycle as primary output after completing sequencing. bcl2fastq Conversion Software (bcl2fastq) combines these per-cycle *.bcl files from a run and translates them into FASTQ files. During the process, bcl2fastq can also remove indexes you used in the sequence. FASTQ file includes sequencing reads and its quality scores which allow you to check base calling errors, poor quality, and adaptor. Similar with RNA-seq data analysis, FASTQC and PRINSEQ can also be used to assess data quality and trim sequence reads for DNA-seq data.
>
> **Step 2:** *Read alignment into reference genome.* The different sequence technologies and the resultant different sequencing characters such as short read with no gap and long reads with gaps have spurred the development of different aligning tools or programs. These include Mapping and Assembly with Qualities (MAQ), Efficient Large-Scale Alignment of Nucleotide Databases (Eland), Bowtie,

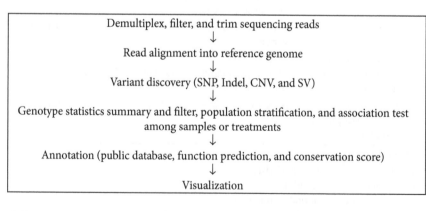

FIGURE 4.1 Genome-seq analysis pipeline.

Short Oligonucleotide Analysis Package (SOAP), and Burrows–Wheeler Aligner (BWA). MAQ is a program that rapidly aligns short sequencing reads to reference genomes. It is particularly designed for Illumina-Solexa 1G Genetic Analyzer. At the mapping stage, MAQ supports ungapped alignment. BWA tool maps low-divergent sequences against a large reference genome. It designs for 70 bp to 1 Mbp reads alignment from Illumina and AB SOLiD sequencing machines. It needs building reference genome and index before alignment, which allow efficient random access to reference genome, and produces alignment in the standard Sequence Alignment/Map (SAM) format. SAM is converted into the Binary Version of SAM (BAM). WGS and WES studies always need more accurately BAM files since raw alignment includes some biased and sequencing errors. Like during library preparation, sequencing errors will be propagated in duplicates when multiple reads come from the same template. Edges of Indels often map with mismatching bases that are mapping artifacts. Base quality scores provided by sequencing machine are inaccurate and biased. Those will affect variation calling. Removing duplicates, local realignment around Indels, and base quality score recalibration are common practices before variation calling. SAM tool provides various utilities for manipulating alignment including sorting, indexing, and merging. Picard is a set of command-line tools for manipulating high-throughput sequencing data and format. Genome Analysis Toolkit (GATK) provides many tools for sequence data processing and variant discovery, variant evaluation, and manipulation. In this chapter, we present the combination of these tools to preprocess the BAM file before variants discoveries.

Step 3: *Variants discovery.* Based on the BAM alignment file, most of variants can be called. Tools for discovering whole genome-wide and whole exome variants can be grouped into SNP, Indel, CNV, and SV identification tools. Among them, SNP and Indel discovery are common. SNP and Indel variants discovery can be divided into two stages: variant calling and post-variant calling filtering. GATK provides a set of tools for the whole pipeline. Unified Genotyper and Haplotype Caller both can call variants. The difference is that Haplotype Caller is a more sophisticated tool and Unified Genotyper can deal with non-diploid organism. Variants filtering also include two methods: Variant Recalibrate and Variant Filtration. Variant Recalibrate uses the machine learning method to train known public variants for recalibrating variants. Variant Filtration uses the fixed thresholds for filtering variants. If you have diploid and enough depth of coverage variants like our example below, Haplotype Caller and Variant Recalibrate are recommended in your analysis. In addition to these, other softwares can also serve the same purpose. FreeBayes uses Bayesian genetic variant detector to find SNPs, Indels, and complex events (composite insertion and substitution events) smaller than reads length. In the tutorial of this chapter, we also use Galaxy freeBayes as an example. When the alignment BAM file is loaded, it will report a standard variant VCF file. Another important variant discovery is to detect genomic copy number variation and structure variation. VarScan is a software to discover somatic mutation and copy number alteration in cancer by exome sequencing. At first, samtools mpileup uses disease and normal BAM files to generate a pileup file. And then, VarScan copy number will detect copy number variations between disease and normal samples. VarScan copy Caller will adjust for GC content and make preliminary calls. ExomeCNV is a R package software, which uses depth of coverage and B-allele for detecting copy number variation and loss of heterozygosity. GATK depth of coverage will be used to convert BAM file into coverage file. Afterward, ExomeCNV will use paired coverage files (e.g., tumor-normal pair) for copy number variation detections. Copy number variations will be called on each exon and large segments one chromosome at a time. BreakDancer has been used to predict wide-variety of SVs including deletion, insertion, inversion, intrachromosomal translocation, and interchromosomal translocation. BreakDancer takes alignment BAM files as input, and bam2cfg will generate a configure file. Based on configure file, BreakDancerMax will detect the five structure variations in the sample.

Step 4: *Genotype statistics summary and filter, population stratification, and association test among samples or treatments.* When you get your variants, you need to know how many known and novel alleles in your samples and what difference between groups, and how many significant variants in your treatment or samples. SNP & Variation Suite (SVS) is a GUI-based software, which includes SNP analysis, DNA-seq analysis, and RNA-seq analysis. Genotype statistics summary will detect minor and major alleles and compute their frequencies in your groups. To avoid SNVs that are triallelic or low-frequency variants, filter setting is necessary. Filtering parameters include call rate, minor allele frequency, hardy Weinberg equilibrium, and number of alleles. After you get high-quality variants data, principal component analysis and the Q–Q plot may be applied to detect and adjust population stratification between normal control populations or cohorts and diseased or treatment populations or cohorts. The final list of those significant different variants may be tenable as potential *causative* variants of interest to predict disease susceptibility, severity, and outcome or treatment response.

Step 5: *Variant annotations.* Most of variant annotations are based on various public variant database (e.g., dbSNP, 1000 genome) to identify known and new variants and on different methods to evaluate impacts of different variants on protein function. The polymorphism phenotyping (PolyPhen) and sorting intolerant from tolerant (SIFT) can predict possible effect of an amino acid substitution on protein function based on straightforward physical and amino acid residues conservation in sequence. Based on variant positions, those tools will give each variant a score which stands for damaging level. ANNOVAR is a command-line tool to annotate functional genetic variants, which includes gene-based annotation, region-based annotations, filter-based annotation, and TABLE_ANNOVAR. Gene-based annotation uses gene annotation system (e.g., UCSC genes and ENSEMBL genes) to identify whether SNPs or CNVs cause protein coding change. Region-based annotations use species-conserved regions to identify variants in specific genomic regions and use transcription factor binding sites, methylation patterns, segmental duplication regions, and so on to annotate variants on genomic intervals. Filter-based annotation uses different public database (dbSNP and 1000 genome) to filter common and rare variants and uses non-synonymous SNPs damaging score like SIFT score and PolyPhen score to identify functional variants. TABLE_ANNOVAR will generate a table file with summary of annotation variants, like gene annotation, amino acid change annotation, SIFT scores, and PolyPhen scores.

Step 6: *Visualization.* The variant visualization step is intended to display how many variants occur in each sample, where they locate, what different structures they potentially engender when comparing with reference genome or other public variants database. Similar with RNA-seq data, Integrative Genomics Viewer (IGV) also can display BAM, VCF, SNP, LOH, and SEG format for location and coverage information of variants. IGV can also reveal the relationship between variants and annotation (e.g., exon, intron, or intergenic). It can also upload GWAS format data that contain *p*-value of the association to display a Manhattan plot. Circos is another command-line tool for visualizing variants' relationship between multiple genome, sequence conservation, and synteny. Meantime, it can display SNP, Indel, CNV, SV, and gene annotation in the same figure.

4.4 STEP-BY-STEP TUTORIAL OF GENOME-SEQ DATA ANALYSIS

Many genome-seq analysis software and pipelines have been developed. Here, we pick GATK pipeline as a command-line interface-based example and Galaxy platform as a GUI-based example.

4.4.1 Tutorial 1: GATK Pipeline

GATK is a software package developed at the Broad Institute to analyze high-throughput sequencing data. The toolkit offers a wide variety of tools, with a primary focus on variant discovery and genotyping as well as strong emphasis on data quality assurance. Here, we use one individual sequencing data (HG01286) from human with 1000 genomes as an example, which was obtained by single-end read sequencing using Illumina's NGS instrument.

Step 1: To download sra data and convert into FASTQ

```
--------------------------------------------------------------------------------
# download SRR1607270.sra data from NCBI FTP service
$wget ftp://ftp-trace.ncbi.nlm.nih.gov/sra/sra-
instant/reads/ByRun/sra/SRR/SRR160/SRR1607270/
SRR1607270.sra
# covert sra format into fastq format
$fastq-dump SRR1607270.sra
# when it is finished, you can check all files:
$ ls -l
# SRR1607270.fastq will be produced.
--------------------------------------------------------------------------------
```

Step 2: To download human genome data and variation annotation files

```
-----------------------------------------------------------------------
# download those data from GATKbundle FTP service
$wget ftp://gsapubftp-anonymous@ftp.broadinstitute.
org/bundle/2.8/hg19/ucsc.hg19.fasta.gz
$wget ftp://gsapubftp-anonymous@ftp.broadinstitute.
org/bundle/2.8/hg19/ucsc.hg19.dict.gz
$wget ftp://gsapubftp-anonymous@ftp.broadinstitute.
org/bundle/2.8/hg19/ucsc.hg19.fasta.fai.gz
$wget ftp://gsapubftp-anonymous@ftp.broadinstitute.
org/bundle/2.8/hg19/1000G_omni2.5.hg19.sites.vcf.gz
$wget ftp://gsapubftp-anonymous@ftp.broadinstitute.
org/bundle/2.8/hg19/1000G_phase1.snps.high_
confidence.hg19.sites.vcf.gz
$wget ftp://gsapubftp-anonymous@ftp.broadinstitute.
org/bundle/2.8/hg19/dbsnp_138.hg19.vcf.gz
$wget ftp://gsapubftp-anonymous@ftp.broadinstitute.
org/bundle/2.8/hg19/hapmap_3.3.hg19.sites.vcf.gz
$wget ftp://gsapubftp-anonymous@ftp.broadinstitute.
org/bundle/2.8/hg19/Mills_and_1000G_gold_standard.
indels.hg19.sites.vcf.gz
# gunzip .gz files
$gunzip *.gz
# when it is finished, you can check all files:
$ ls -l
# ucsc.hg19.fasta, ucsc.hg19.dict, ucsc.hg19.fasta.
fai, 1000G_omni2.5.hg19.sites.vcf, 1000G_phase1.snps.
high_confidence.hg19.sites.vcf , dbsnp_138.hg19.vcf,
hapmap_3.3.hg19.sites.vcfand Mills_and_1000G_gold_
standard.indels.hg19.sites.vcf will be produced.
-----------------------------------------------------------------------
```

Step 3: To index human genome

```
-----------------------------------------------------------------------
```

BWA index will be used to build genome index which allows efficient random access to the genome before reads alignment.

```
-----------------------------------------------------------------------
# generate BWA human index
$bwa index ucsc.hg19.fasta
```

```
# when it is finished, you can check all file:
$ ls -l
# ucsc.hg19.fasta.amb, ucsc.hg19.fasta.ann, ucsc.
hg19.fasta.bwt, ucsc.hg19.fasta.pac and ucsc.hg19.
fasta.sa will be produced.
```

Step 4: To map single-end reads into reference genome

Burrows–Wheeler Aligner (BWA) maps sequencing reads against reference genome. There are three aligning or mapping algorithms designed for Illumina sequence reads from 70 bp to 1 Mbp. Here, BWA-MEM will align fastq files (SRR1607270.fastq) into human UCSC hg19 genome (ucsc.hg19.fasta). The generated SAM file contains aligning reads.

```
$bwa mem ucsc.hg19.fasta SRR1607270.fastq >sample.
sam
# when it is finished, you can check all files:
$ ls -l
# sample.sam will be produced.
```

Step 5: To sort SAM into BAM

Picard SortSam is used to convert SAM file (sample.sam) into BAM file (sample.bam), and sort BAM file order by starting positions.

```
$ java -jar /data/software/picard/SortSam.jar
INPUT=sample.sam OUTPUT=sample.bam
SORT_ORDER=coordinate
# when it is finished, you can check file:
$ ls -l
# sample.bam will be produced.
```

Step 6: To mark duplicate reads

During the sequencing process, the same sequences can be sequenced several times. When sequencing error appears, it will be propagated in duplicates. Picard MarkDuplicates is used to flag read duplicates. Here, input file is sample.bam which was coordinate sorted by picard-SortSam, output file is sample_dedup.bam file which contains marked duplicated reads, duplication metrics will be written in metrics.txt.

```
$ java -jar /data/software/picard/MarkDuplicates.jar
INPUT=sample.bam OUTPUT=sample_dedup.bam METRICS_
FILE=metrics.txt
# when it is finished, you can check all files:
$ ls -l
# sample_dedup.bam and metrics.txt will be produced.
```

Step 7: To add read group information

The read group information is very important for downstream GATK functionality. Without a read group information, GATK will not work. Picard AddOrReplaceReadGroups replaces all read groups in the input file (sample_dedup.bam) with a single new read group and assigns all reads to this read group in the output BAM (sample_AddOrReplaceReadGroups.bam). Read group library (RGLB), read group platform (RGPL), read group platform unit (RGPU), and read group sample name (RGSM) will be required.

```
$ java -jar /data/software/picard/
AddOrReplaceReadGroups.jar RGLB=L001 RGPL=illumina
RGPU=C2U2AACXX RGSM=Sample I=sample_dedup.bam
O=sample_AddOrReplaceReadGroups.bam
# when it is finished, you can check all file:
$ ls -l
# sample_AddOrReplaceReadGroups.bam will be produced.
```

Step 8: To index BAM file

Samtools index bam file (sample_AddOrReplaceReadGroups.bam) is for fast random access to reference genome. Index file sample_AddOrReplaceReadGroups.bai will be created.

```
$ samtools index sample_AddOrReplaceReadGroups.bam
# when it is finished, you can check all files:
$ ls -l
# sample_AddOrReplaceReadGroups.bai will be produced.
```

Step 9: To realign locally around Indels

Alignment artifacts result in many bases mismatching the reference near the misalignment, which are easily mistaken as SNPs. Realignment around Indels helps improve the accuracy. It takes two steps: GATK RealignerTargetCreator firstly identifies what regions need to be realigned and then GATK IndelRealigner performs the actual realignment. Here, Mills_and_1000G_gold_standard.indels.hg19.sites.vcf is used as known Indels for realignment and UCSC hg19 (ucsc.hg19.fasta) is used as reference genome. Output (sample_realigner.intervals) will contain the list of intervals identified as needing realignment for IndelRealigner, and output (sample_realigned.bam) will contain all reads with better local alignments.

```
$java -jar /data/software/gatk-3.3/GenomeAnalysisTK.
jar -T RealignerTargetCreator -R ucsc.hg19.fasta -I
sample_AddOrReplaceReadGroups.bam --known Mills_
and_1000G_gold_standard.indels.hg19.sites.vcf -o
sample_realigner.intervals
$java -jar /data/software/gatk-3.3/GenomeAnalysisTK.
jar -I sample_AddOrReplaceReadGroups.bam -R ucsc.
hg19.fasta -T IndelRealigner -targetIntervals
sample_realigner.intervals -known Mills_and_1000G_
```

```
gold_standard.indels.hg19.sites.vcf -o sample_
realigned.bam
# when it is finished, you can check all files:
$ ls -l
#sample_realigner.intervals and sample_realigned.bam
will be produced.
```

Step 10: To recalibrate base quality score

Due quality scores accessed by sequencing machines are inaccurate or biased, recalibration of base quality score is very important for downstream analysis. The recalibration process divides into two steps: GATK BaseRecalibrator models an empirically accurate error model to recalibrate the bases and GATK PrintReads applies recalibration to your sequencing data. The known sites (dbsnp_138.hg19.vcf and Mills_and_1000G_gold_standard.indels.hg19.sites.vcf) are used to build the covariation model and estimate empirical base qualities. The output file sample_BaseRecalibrator.grp contains the covariation data to recalibrate the base qualities of your sequence data. Output file sample_PrintReads.bam will list reads with accurate base substitution, insertion and deletion quality scores.

```
$ java -jar /data/software/gatk-3.3/
GenomeAnalysisTK.jar -I sample_realigned.bam -R
ucsc.hg19.fasta -T BaseRecalibrator -known Sites
dbsnp_138.hg19.vcf -knownSites Mills_and_1000G_gold_
standard.indels.hg19.sites.vcf -o sample_
BaseRecalibrator.grp
$ java -jar /data/software/gatk-3.3/
GenomeAnalysisTK.jar -R ucsc.hg19.fasta -T
PrintReads -BQSR sample_BaseRecalibrator.grp -I
sample_realigned.bam-o sample_PrintReads.bam
# when it is finished, you can check all files:
$ ls -l
# sample_BaseRecalibrator.grp and sample_PrintReads.
bam will be produced.
```

Step 11: To call variant

HaplotypeCaller can call SNPs and Indels simultaneously via a local de-novo assembly. It will convert alignment bam file (sample1_ PrintReads.bam) into variant call format VCF file (raw_sample.vcf).

```
$ java -jar /data/software/gatk-3.3/
GenomeAnalysisTK.jar -T HaplotypeCaller -ERC GVCF
-variant_index_type LINEAR -variant_index_parameter
128000 -R ucsc.hg19.
fasta -I sample_PrintReads.bam -stand_emit_conf 10
-stand_call_conf 30 -o raw_sample.vcf
# when it is finished, you can check all files:
$ ls -l
# raw_sample.vcf will be produced.
```

Step 12: To recalibrate variant quality scores for SNPs

When you get high sensitivity raw callsets, you need to recalibrate variant quality scores to filter raw variations, further reduce parts of false positives. Due to different character of SNPs and Indels, you will separate SNPs and Indels to recalibrate variant quality scores. GATK VariantRecalibrator applies machine learning method which use hap-map, omin, dbSNP, and 1000 high-confidence variants as known/true SNP variants for training model, and then use the model to recalibrate our data. GATK ApplyRecalibration applies the recalibration lever to filter our data. Output file sample_recalibrate_SNP.recal will contain recalibrated data, output file sample_recalibrate_SNP.tranches will contain quality score thresholds, and output file sample_recal.SNPs.vcf will contain all SNPs with recalibrated quality scores and flag PASS or FILTER.

```
$ java -jar /data/software/gatk-3.3/
GenomeAnalysisTK.jar -T VariantRecalibrator -R ucsc.
hg19.fasta -input raw_sample.vcf -resource:hapmap,kn
own=false,training=true,truth=true,prior=15.0
```

```
hapmap_3.3.hg19.sites.vcf -resource:omni,known=
false,training=true,truth=false,prior=12.0 1000G_
omni2.5.hg19.sites.vcf -resource:1000G,known=false,
training=true,truth=false,prior=10.0 1000G_phase1.
snps.high_confidence.hg19.sites.vcf -resource:dbsnp,
known=true,training=false,truth=false,prior=6.0
dbsnp_138.hg19.vcf -an QD -an MQ -an MQRankSum -an
ReadPosRankSum -an FS -mode SNP -recalFile sample_
output.recal -tranchesFile sample_recalibrate_SNP.
tranches -rscriptFile sample_recalibrate_SNP_plots.R
$ java -jar /data/software/gatk-3.3/
GenomeAnalysisTK.jar -T ApplyRecalibration -R ucsc.
hg19.fasta -input raw_sample.vcf -mode SNP
-recalFile sample_recalibrate_SNP.recal
-tranchesFile sample_recalibrate_SNP.tranches -o
sample_recal.SNPs.vcf --ts_filter_level 99.0
# when it is finished, you can check all files:
$ ls -l
# sample_recalibrate_SNP.recal, sample_recalibrate_
SNP.tranches and sample_recal.SNPs.vcf will be
produced.
```

Step 13: To recalibrate variant quality scores for Indels

Same process with recalibration variant quality scores of SNPs, GATK VariantRecalibrator, and ApplyRecalibration will be used to recalibrate Indels. Mills_and_1000G_gold_standard.indels. hg19.sites.vcf will be used to train Indels model. Finally, Output file sample_final_recalibrated_variants.vcf will contain all SNPs and Indels with recalibrated quality scores and flag PASS or FILTER.

```
$ java -jar /data/software/gatk-3.3/
GenomeAnalysisTK.jar -T VariantRecalibrator -R
ucsc.hg19.fasta -input sample_recal.SNPs.vcf -resou
rce:mills,known=true,training=true,truth=true,pr
ior=12.0 Mills_and_1000G_gold_standard.indels.hg19.
sites.vcf -an DP -an FS -an MQRankSum -an
ReadPosRankSum -mode INDEL -minNumBad 1000
```

```
--maxGaussians 4 -recalFile sample_recalibrate_
INDEL.recal -tranchesFile sample_recalibrate_INDEL.
tranches -rscriptFile
sample_recalibrate_INDEL_plots.R
$ java -jar /data/software/gatk-3.3/
GenomeAnalysisTK.jar-T ApplyRecalibration -R ucsc.
hg19.fasta -input sample_recal.SNPs.vcf -mode INDEL
--ts_filter_level 99.0 -recalFile sample_
recalibrate_INDEL.recal -tranchesFile sample_
recalibrate_INDEL.tranches -o
sample_final_recalibrated_variants.vcf
# when it is finished, you can check all files:
$ ls -l
# sample_recalibrate_INDEL.recal, sample_
recalibrate_INDEL.tranches and sample_final_
recalibrated_variants.vcf will be produced.
```

Note:

1. $ is a prompt sign for command or command-line input for each step.

2. # indicates a comment for each step.

More details can be found in https://www.broadinstitute.org/gatk/guide/tooldocs/

4.4.2 Tutorial 2: Galaxy Pipeline

Galaxy is an open-source, web-based platform for data intensive biomedical research. Galaxy supplies many tools for variants detection of genome-seq data such as FreeBayes, GATK, VarScan, ANNOVAR, snpEff. Here, we provide an example that shows you how to analyze raw fastq file to obtain variation call and annotation in the galaxy.

> **Step 1:** *Transfer SRR1607270.fastq data into Galaxy FTP server.* If your file size is bigger than 2GB, you need to upload your data via FTP. At first, download and install **FileZilla** in your computer. Then open **FileZilla**, set Host "**usegalaxy.org**," your **Username** and **Password**, click **Quickconnect**. Select your file SRR1607270.fastq from your

local site and drag your file into blank area in the remote site. The status of file transfer process can be followed on the screen. When it is finished, you can continue to the next step.

Step 2: *Upload SRR1607270.fastq via the Galaxy FTP server.* Open https://usegalaxy.org/ and **login in.** Click **Get Data -> Upload File from your computer**, then click **Choose FTP file** -> click **Start.**

Step 3: *Edit SRR1607270.fastq attributes.* Click on **pencil icon** adjacent to SRR1607270.fastq in History windowS, then click **Datatype** and select **fastqillumina**, click **Attributes** and select Database/Build **Human Feb.2009 (GRCh37/hg19) (hg19)** as reference, and click **Save.**

Step 4: *Report quality of SRR1607270.fastq.* Click **QC and manipulation -> FastQC: Read QC reports using FastQC**, then select **SRR1607270.fastq** and click **Execute.** The process and result will appear in **History** window.

Step 5: *Map SRR1607270.fastq into human genome.* Click **NGS: Mapping -> Map with BWA for Illumina**, and then chose different parameters for alignment. Here, you can select **Use a built-in index** and **Human (homo sapiens) (b37): hg19** as reference genome and index, **Single-end** as library type, FASTQ file **SRR1607270.fastq**, and BWA settings **Commonly used**, and click **Execute.** When it is finished, bam file will appear in **History** window.

Step 6: *Call variants.* Click **NGS: Variant Analysis -> FreeBayes**, select **hg19** as reference genome, choose **1: Simple diploid calling** as parameter selection level, and **Execute.** When it is finished, vcf file including all variants will appear in **History** window. You can click the result, it will show all details and you also can download the vcf file into your computer.

Step 7: *Variant annotation.* Click **NGS: Variant Analysis -> ANNOVAR Annotate VCF**, select **FreeBayes variants** as variants file, choose **refGene** as gene annotations, **phastConsElements46way** as annotation regions and **1000g2012apr_all** as annotation databases parameters, click **Execute.** The results will appear in **History** window, which will show you protein coding change and amino acids affect.

REFERENCES

Bansal V, Bashir A, Bafna V: Evidence for large inversion polymorphisms in the human genome from HapMap data. *Genome Research* 2007, **17**(2):219–230.

Chen K, Wallis JW, McLellan MD, Larson DE, Kalicki JM, Pohl CS, McGrath SD, Wendl MC, Zhang Q, Locke DP et al.: BreakDancer: An algorithm for high-resolution mapping of genomic structural variation. *Nature Methods* 2009, **6**(9):677–681.

Cheranova D, Zhang LQ, Heruth D, Ye SQ: Chapter 6: Application of next-generation DNA sequencing in medical discovery. In *Bioinformatics: Genome Bioinformatics and Computational Biology*. 1st ed., pp. 123–136, (eds) Tuteja R, Nova Science Publishers, Hauppauge, NY, 2012.

Chmielecki J, Crago AM, Rosenberg M, O'Connor R, Walker SR, Ambrogio L, Auclair D, McKenna A, Heinrich MC, Frank DA et al.: Whole-exome sequencing identifies a recurrent NAB2-STAT6 fusion in solitary fibrous tumors. *Nature Genetics* 2013, **45**(2):131–132.

Fairfax BP, Humburg P, Makino S, Naranbhai V, Wong D, Lau E, Jostins L, Plant K, Andrews R, McGee C et al.: Innate immune activity conditions the effect of regulatory variants upon monocyte gene expression. *Science* 2014, **343**(6175):1246949.

Genome of the Netherlands C: Whole-genome sequence variation, population structure and demographic history of the Dutch population. *Nature Genetics* 2014, **46**(8):818–825.

Genomes Project C, Abecasis GR, Altshuler D, Auton A, Brooks LD, Durbin RM, Gibbs RA, Hurles ME, McVean GA: A map of human genome variation from population-scale sequencing. *Nature* 2010, **467**(7319):1061–1073.

Genomes Project C, Abecasis GR, Auton A, Brooks LD, DePristo MA, Durbin RM, Handsaker RE, Kang HM, Marth GT, McVean GA et al.: An integrated map of genetic variation from 1,092 human genomes. *Nature* 2012, **491**(7422):56–65.

Lee E, Iskow R, Yang L, Gokcumen O, Haseley P, Luquette LJ, 3rd, Lohr JG, Harris CC, Ding L, Wilson RK et al.: Landscape of somatic retrotransposition in human cancers. *Science* 2012, **337**(6097):967–971.

MacArthur DG, Balasubramanian S, Frankish A, Huang N, Morris J, Walter K, Jostins L, Habegger L, Pickrell JK, Montgomery SB et al.: A systematic survey of loss-of-function variants in human protein-coding genes. *Science* 2012, **335**(6070):823–828.

Priebe L, Degenhardt FA, Herms S, Haenisch B, Mattheisen M, Nieratschker V, Weingarten M, Witt S, Breuer R, Paul T et al.: Genome-wide survey implicates the influence of copy number variants (CNVs) in the development of early-onset bipolar disorder. *Molecular Psychiatry* 2012, **17**(4):421–432.

Sathirapongsasuti JF, Lee H, Horst BA, Brunner G, Cochran AJ, Binder S, Quackenbush J, Nelson SF: Exome sequencing-based copy-number variation and loss of heterozygosity detection: ExomeCNV. *Bioinformatics* 2011, **27**(19):2648–2654.

Soden SE, Saunders CJ, Willig LK, Farrow EG, Smith LD, Petrikin JE, LePichon JB, Miller NA, Thiffault I, Dinwiddie DL et al.: Effectiveness of exome and genome sequencing guided by acuity of illness for diagnosis of neurodevelopmental disorders. *Science Translational Medicine* 2014, **6**(265):265ra168.

Zack TI, Schumacher SE, Carter SL, Cherniack AD, Saksena G, Tabak B, Lawrence MS, Zhang CZ, Wala J, Mermel CH et al.: Pan-cancer patterns of somatic copy number alteration. *Nature Genetics* 2013, **45**(10):1134–1140.

## CHAPTER 5

RNA-Seq Data Analysis

Li Qin Zhang, Min Xiong, Daniel P. Heruth, and Shui Qing Ye

CONTENTS

5.1 INTRODUCTION

RNA-sequencing (RNA-seq) is a technology that uses next-generation sequencing (NGS) to determine the identity and abundance of all RNA sequences in biological samples. RNA-seq is gradually replacing DNA microarrays as a preferred method for transcriptome analysis because it has the advantages of profiling a complete transcriptome, not relying on any known genomic sequence, achieving *digital* transcript expression analysis with a potentially unlimited dynamic range, revealing sequence variations (single-nucleotide polymorphisms [SNPs], fusion genes, and isoforms) and providing allele-specific or isoform-specific gene expression detections.

RNA is one of the essential macromolecules in life. It carries out a broad range of functions, from translating genetic information into the molecular machines and structures of the cell by mRNAs, tRNAs, rRNAs, and others to regulating the activity of genes by miRNAs, siRNAs, lincRNAs, and others during development, cellular differentiation, and changing environments. The characterization of gene expression in cells via measurement of RNA levels with RNA-seq is frequently employed to determine how the transcriptional machinery of the cell is affected by external signals (e.g., drug treatment) or how cells differ between a healthy state and a diseased state. RNA expression levels often correlate with functional roles of their cognate genes. Some molecular features can only be observed at the RNA level such as alternative isoforms, fusion transcripts, RNA editing, and allele-specific expression. Only 1%–3% RNAs are protein coding RNAs, while more than 70% RNAs are non-coding RNAs. Their regulatory roles or other potential functions may only be gleaned by analyzing the presence and abundance of their RNA expressions.

A number of NGS platforms for RNA-seq and other applications have been developed. Several major NGS platforms are briefed here. Illumina's platform (http://www.illumina.com/) represents one of the most popularly used sequencing by synthesis chemistry in a massively parallel arrangement. Currently, it markets HiSeq X Five and HiSeq X Ten instruments with population power; HiSeq 2500, HiSeq 3000, and HiSeq 4000 instruments with production power; Nextseq 500 with flexible power; and MiSeq with focused power. The HiSeq X Ten is a set of 10 ultra-high-throughput sequencers, purpose-built for large-scale human whole-genome sequencing at a cost of $1000 per genome, which together can sequence over 18,000 genomes per year. The MiSeq desktop sequencer allows you to access more focused applications such as targeted gene sequencing, metagenomics, small-genome sequencing, targeted gene expression, amplicon sequencing, and HLA typing. New MiSeq reagents enable up to 15 GB of output with 25 M sequencing reads and 2 × 300 bp read lengths. Life Technologies (http://www.lifetechnologies.com/) markets sequencing by oligonucleotide ligation and detection (SOLID) 5500 W Series Genetic Analyzers, Ion Proton™ System, and the Ion Torrent™ Personal Genome Machine® (Ion PGM™) System. The newest 5500 W instrument uses flow chips, instead of beads, to amplify templates, thus simplifying the workflow and reducing costs. Its sequencing accuracy can be up to 99.99%. The Ion Proton™ and Ion PGM™ Systems are ion semiconductor-based platforms. The Ion PGM™ System is one of top selling benchtop NGS solutions. Roche markets 454 NGS platforms (http://www.454.com/), the GS FLX+ System, and the GS Junior Plus System. They are based on sequencing by synthesis chemistry. The GS FLX+ System features the unique combination of long reads (up to 1000 bp), exceptional accuracy and high-throughput, making the system well suited for larger genomic projects. The GS Junior Plus System is a benchtop NGS platform suitable for individual lab NGS needs. Pacific Biosciences (http://www.pacificbiosciences.com/) markets the PACBIO RSII platform. It is considered as the third-generation sequencing platform since it only requires a single molecule and reads the added nucleotides in real time. The chemistry has been termed SMRT for single-molecule real time. The PacBio RS II sequencing provides average read lengths in excess of >10 KB with ultra-long reads >40 KB. The long reads are characterized by high 99.999% consensus accuracy and are ideal for de novo assembly, targeted sequencing applications, scaffolding, and spanning structural rearrangements. Oxford Nanopore Technologies (https://nanoporetech.com/) markets the GridION™ system, the PromethION, and the MinION™ devices. Nanopore sequencing is a third-generation single-molecule technique. The GridION™ system is a benchtop instrument and an electronics-based platform. This enables multiple nanopores to be measured simultaneously and data to be sensed, processed, and analyzed in real time. The PromethION is a tablet-sized benchtop instrument designed to run a small number of samples. The MinION device is a miniaturized single-molecule analysis system, designed for single use and to work through the USB port of a laptop or desktop computer. With continuous improvements and refinements, nanopore-based sequencing technology may gain its market share in not distant future.

5.2 RNA-SEQ APPLICATIONS

RNA-seq is commonly applied to identify the sequence, structure, and abundance of RNA molecules in a specific sample. The nature of questions one may address using RNA-seq technology is effectively limitless, and thus, it is virtually impossible to present an exhaustive list of all current and potential RNA-seq applications. Table 5.1 lists some representative applications of RNA-seq.

TABLE 5.1   RNA-Seq Applications

#	Usages	Descriptions	References
1	Differential gene expression analysis	Comparing the abundance of RNAs among different samples	Wang et al. (2009)
2	Transcript annotations	Detecting novel transcribed regions, splice events, additional promoters, exons, or untranscribed regions	Zhou et al. (2010), Mortazavi et al. (2008)
3	ncRNA profiling	Identifying non-coding RNAs (lncRNAs, miRNAs, siRNAs, piRNAs, etc.)	Ilott et al. (2013)
4	eQTL[a]	Correlating gene expression data with known SNPs	Majewski et al. (2011)
5	Allele-specific expression	Detecting allele-specific expression	Degner et al. (2009)
6	Fusion gene detection	Identification of fusion transcripts	Edgren et al. (2011)
7	Coding SNP discovery	Identification of coding SNPs	Quinn et al. (2013)
8	Repeated elements	Discovery of transcriptional activity in Repeated elements	Cloonan et al. (2008)
9	sQTL[b]	Correlating splice site SNPs with gene expression levels	Lalonde et al. (2011)
10	Single-cell RNA-seq	Sequencing all RNAs from a single cell	Hashimshony et al. (2012)
11	RNA-binding site identification	Identifying RNA-binding sites of RNA Binding proteins using CLIP-seq[c], PAR-CLIP[d], and iCLIP[e]	Darnell et al. (2010) Hafner et al. (2010) Konig et al. (2010)
12	RNA-editing site identification	Identifying RNA-editing sites	Ramaswami et al. (2013) Danecek et al. (2012)

[a] eQTL, expression quantitative trait loci.
[b] sQTL, splice site quantitative trait loci.
[c] CLIP-seq, cross-linking immunoprecipitation sequencing.
[d] PAR-CLIP, photoactivatable-ribonucleoside-enhanced CLIP.
[e] iCLIP, individual-nucleotide resolution CLIP.

5.3 RNA-SEQ DATA ANALYSIS OUTLINE

Data analysis is perhaps the most daunting task of RNA-seq. The continued improvement in sequencing technologies have allowed for the acquisition of millions of reads per sample. The sheer volume of these data can be intimidating. Similar to advances in the sequencing technology, there have been continued development and enhancement in software packages for RNA-seq analysis, thus providing more accessible and user friendly bioinformatic tools. Because the list of common and novel RNA-seq applications is growing daily and there are even more facets to the analysis of RNA-seq data than there are to generating the data itself, it would be difficult, if not impossible, to cover all developments in approaches to analyzing RNA-seq data. The objective of this section is to provide general outline to commonly encountered steps and questions one faces on the path from raw RNA-seq data to biological conclusion. Figure 5.1 provides example workflow, which assumes that a reference genome is available.

Step 1: *Demultiplex, filter, and trim sequencing reads.* Many researchers multiplex molecular sequencing libraries derived from several samples into a single pool of molecules to save costs because of a high sequence output from a powerful next-generation sequencer, such as Illumina 2500, more than the coverage need of the RNA-seq of a single sample. Multiplexing of samples is made possible by incorporation of a short (usually at least 6 nt) index or *barcode* into each DNA fragment during the adapter ligation or PCR amplification steps of library preparation. After sequencing, each read can be traced back to its original sample using the index sequence and binned accordingly. In the case of Illumina sequencing, barcodes that are variable across samples at the first few bases are used to ensure adequate cluster discrimination. Many programs have been written to demultiplex barcoded library pools. Illumina's software bcl2fastq2 Conversion Software (v2.17) can demultiplex multiplexed samples during the step converting *.bcl files into *.fastq.gz files (compressed FASTQ

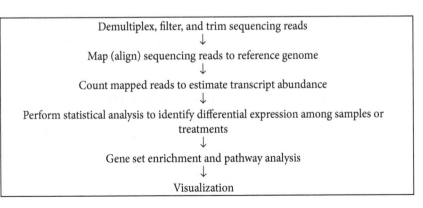

FIGURE 5.1   RNA-seq data analysis pipeline. See text for a brief explanation of each step.

files). bcl2fastq2(v2.17) can also align samples to a reference sequence using the compressed FASTQ files and call SNPs and indels, and perform read counting for RNA sequences. Quality control of raw sequencing data by filter and trim is usually carried out before they will be subjected to the downstream analysis. Raw sequencing data may include low confidence bases, sequencing-specific bias, 3'/5' position bias, PCR artifacts, untrimmed adapters, and sequence contamination. Raw sequence data are filtered by the real-time analysis (RTA) software to remove any reads that do not meet the overall quality as measured by the Illumina chastity filter, which is based on the ratio of the brightest intensity divided by the sum of the brightest and second brightest intensities. The default Illumina pipeline quality filter threshold of passing filter is set at CHASTITY $\geq$ 0.6, that is, no more than one base call in the first 25 cycles has a chastity of <0.6. A few popular filter and trim software are noted here. FastQC (http://www.bioinformatics.bbsrc.ac.uk/projects/fastqc/) can provide a simple way to do some quality control checks on raw sequence data. PRINSEQ (http://prinseq.sourceforge.net/) can filter, trim, and reformat NGS data. In PRINSEQ, you can combine many trimming and filtering options in one command. Trimomatic (http://www.usadellab.org/cms/?page=trimmomatic) can perform a variety of useful trimming tasks for NGS data.

Step 2: *Align sequencing reads to reference genome.* The goal of this step is to find the point of origin for each and every sequence read. Both new sequence data and a reference sequence are needed to align the former to the latter. Aligning is computationally demanding because there are millions of reads to align and most reference genomes are large. The genome sequence is often transformed and compressed into an index to speed up aligning. The most common one in use is the Burrows–Wheeler transform. The relatively fast and memory efficient TopHat (http://ccb.jhu.edu/software/tophat/index.shtml) is a commonly used spliced alignment program for RNA-seq reads. TopHat 2.0.14, its current version as of March 24, 2015, aligns RNA-seq reads to mammalian-sized genomes using the ultra-high-throughput short-read aligner Bowtie (http://bowtie-bio.sourceforge.net/index.shtml) and then analyzes the mapping results to identify splice junctions between exons. Bowtie indexes the genome with a Burrows–Wheeler index to keep its memory footprint small: typically about 2.2 GB for the human genome (2.9 GB for paired-end). For the file format of human reference sequence, TopHat needs the annotation file GRCh38.78.gtf, which can be downloaded from the Ensembl site (http://uswest.ensembl.org/info/data/ftp/index.html), and the Bowtie2 (Version 2.2.5—3/9/2015) genome index to build a Bowtie2 transcriptome index which has the basename GRCh38.78.tr. The chromosome names in the gene and transcript models (GTF) file and the genome index must match. Bowtie2 has to be on the path because TopHat2 uses it to build the index. TopHat2 accepts both FASTQ and FASTA file formats of newly generated sequence files as input. The output from this step is an alignment file, which lists the mapped reads and their mapping positions in the reference. The output is usually in a BAM (.bam) file format, which is the binary version of a SAM file. These files contain all of the information for downstream analyses such as annotation, transcript abundance comparisons, and polymorphism detection. Another aligner, Spliced Transcripts Alignment to Reference (STAR, https://github.com/alex-dobin/STAR) software, is emerging as an ultrafast universal RNA-seq aligner. It was based on a previously undescribed RNA-seq alignment algorithm that uses sequential maximum mappable seed search in uncompressed suffix arrays followed by seed clustering and stitching procedure. STAR can not only increase aligning speed but also improve alignment sensitivity and precision. In addition to unbiased de novo detection of canonical junctions, STAR can discover non-canonical splices and chimeric (fusion) transcripts and is also capable of mapping full-length RNA sequences. In the next section, we will present tutorials using both TopHat2 and Star to align sample mouse RNA-seq data to mouse reference genome or transcriptome.

Steps 3 and 4: *Count mapped reads to estimate transcript abundance and perform statistical analysis to identify differential expression among samples or treatments.* A widely adopted software suite, Cufflinks (http://cole-trapnell-lab.github.io/cufflinks/announcements/cufflinks-github/) (Version 2.2.1 5052014) can perform transcriptome assembly and estimate transcript abundance and differential expression analysis for RNA-seq. Cufflinks is the name of a suite of tools which include several programs: Cufflinks, Cuffdiff, Cuffnorm, Cuffmerge, Cuffcompare, and Cuffquant. Cufflinks and Cuffdiff are two most frequently used programs in the Cufflinks suite. Cufflinks assembles transcriptomes from RNA-seq data and quantifies their expression. It takes a text file of SAM alignments or a binary SAM (BAM) file as input. Cufflinks produces three output files: transcriptome assembly: transcripts.gtf; transcript-level expression: isoforms. fpkm_tracking; and gene-level expression: genes.fpkm_tracking. The values of both transcripts and genes are reported as FPKM (fragment per thousand nucleotide per million mapped reads). Cuffdiff is a highly accurate tool for comparing expression levels of genes and transcripts in RNA-seq experiments between two or more conditions as well as for reporting which genes are differentially spliced or are undergoing other types of isoform-level regulation. Cuffdiff takes a GTF2/GFF3 file of transcripts as input, along with two or more BAM or SAM files containing the fragment alignments for two or more samples. It outputs the tab delimited file which lists the results of differential expression testing between samples for spliced transcripts, primary transcripts, genes, and coding sequences. The remaining programs in the Cufflinks suite are optional. Expression levels reported by Cufflinks in FPKM units are usually comparable between samples but in certain situations, applying an extra level of normalization can remove sources of bias in the data. Cuffnorm has two additional normalization method options: the median of the geometric means of fragment counts and the ratio of the 75 quartile fragment counts to the average 75 quartile value across all libraries. It normalizes a set of samples to be on as similar scales as possible, which can improve the results you obtain with other downstream tools. Cuffmerge merges multiple RNA-seq assemblies into a master transcriptome. This step is required for a differential expression analysis of the new transcripts. Cuffcompare can compare the new transcriptome assembly to known transcripts and assess the quality of the new assembly. Cuffquant allows you to compute the gene and transcript expression profiles and save these profiles to files that you can analyze later with Cuffdiff or Cuffnorm. This can help you distribute your computational load over a cluster and is recommended for analyses involving more than a handful of libraries.

Step 5: *Gene set enrichment and pathway analysis.* The output list of differentially expressed genes or transcripts between two or more groups can be shortened by applying different cutoff thresholds, for example, twofold difference and/or p-value $< .01$. One useful way to compare groups of transcripts or genes that are differentially expressed is through gene ontology (GO) term analysis (www.geneontology.org). The terms belong to one of three basic **ontologies**: cellular component, biological process, and molecular function. This analysis can inform the investigator which cellular component, biological process, and molecular function are predominantly dysregulated. QIAGEN'S ingenuity pathway analysis (http://www.ingenuity.com/products/ipa) has been broadly adopted by the life science research community to get a better understanding of the isoform-specific biology resulting from RNA-seq experiments. It unlocks the insights buried in experimental data by quickly identifying relationships, mechanisms, functions, and pathways of relevance. The Database for Annotation, Visualization and Integrated Discovery (DAVID) is a popular free program (http://david.abcc.ncifcrf.gov/), which provides a comprehensive set of functional annotation tools for investigators to understand biological meaning behind large list of differentially expressed genes or transcripts. DAVID currently covers over 40 annotation categories, including GO terms, protein–protein interactions, protein functional domains, disease associations, biopathways, sequence general features, homologies, gene functional summaries, gene tissue expressions, and literatures. DAVID's functional classification tool provides a rapid means to organize large lists of differentially expressed genes or transcripts into functionally related groups to help unravel the biological content captured by high-throughput technologies such as RNA-seq.

Step 6: *Visualization.* It is important to visualize reads and results in a genomic context during the different stages of analysis in order to gain insights into gene and transcript structure and to obtain a sense of abundance. One example is the Integrative Genomics Viewer (IGV, http://www.broadinstitute.org/igv/), which allows one to view the RNA-seq as well as other genomic data. Another example is the CummeRbund (http://compbio.mit.edu/cummeRbund/), which is an R package designed to aid and simplify the task of analyzing Cufflinks RNA-seq output. CummeRbund takes the various output files from a cuffdiff run and creates a SQLite database of the results describing appropriate relationships between genes, transcripts, transcription start sites, and coding sequences (CDS) regions. Once stored and indexed, data for these features, even across multiple samples or conditions, can be retrieved very efficiently and allows the user to explore subfeatures of individual genes, or gene sets as the analysis requires. CummeRbund has implemented numerous plotting functions as well for commonly used visualizations.

5.4 STEP-BY-STEP TUTORIAL ON RNA-SEQ DATA ANALYSIS

There are a plethora of both Unix-based command line and graphical user interface (GUI) software available for RNA-seq data analysis. The open-source, command line Tuxedo Suite, comprised of Bowtie, TopHat, and Cufflinks, has been a popular software suite for RNA-seq data analysis. Due to its both analytical power and ease of use, Tuxedo Suite has been incorporated into several open source and GUI platforms, including Galaxy (galaxyproject.org), Chipster (chipster.csc.fi), GenePattern (http://www.broadinstitute. org/cancer/software/genepattern/), and BaseSpace® (BaseSpace®.illumina. com). In this section, we will demonstrate step-by-step tutorial on two distinct RNA-seq data analysis workflows. First, we will present an Enhanced Tuxedo Suite command line pipeline followed by a review of RNA Express, a GUI workflow available on Illumina's BaseSpace®. Due to the space limitation, gene set enrichment and pathway analysis, as well as the visualization step of final results, will not be demonstrated in this section.

5.4.1 Tutorial 1: Enhanced Tuxedo Suite Command Line Pipeline

Here, we present the command workflow for in-depth analysis of RNA-seq data. Command line-based pipelines typically require a local cluster for both the analysis and storage of data, so you must include these considerations when you plan your RNA-seq experiments. The command line pipeline combines five different tools to do this. MaSuRCA is used to assemble super-reads, TopHat is used to align those reads into genome, StringTie is used to assemble transcripts, Cuffmerge is used to merge two transcriptomes, and Cuffdiff identifies differential expression genes and transcripts between groups. Here, we use two data samples (SRR1686013.sra from decidual stromal cells and SRR1686010.sra from endometrial stromal fibroblasts) of paired-end sequencing reads generated on an Illumina Genome Analyzer II instrument.

Step 1: To download the required programs

----

   a. StringTie (http://ccb.jhu.edu/software/stringtie/)
   b. MaSuRCA(http://www.genome.umd.edu/masurca.html)

c. Cufflinks (http://cole-trapnell-lab.github.io/cufflinks/install/)

d. superreads.pl script (http://ccb.jhu.edu/software/stringtie/dl/superreads.pl)

Step 2: To download sra data and convert into FASTQ

```
# create directories for SRR1686013
$ mkdir SRR1686013
$ cd SRR1686013
$ wgetftp://ftp-trace.ncbi.nlm.nih.gov/sra/sra-
instant/reads/ByRun/sra/SRR/SRR168/SRR1686013/
SRR1686013.sra
$ fastq-dump --split-files SRR1686013.sra

# create directories for SRR1686010
$ mkdir ../SRR1686010
$ cd ../SRR1686010
$ wget ftp://ftp-trace.ncbi.nlm.nih.gov/sra/sra-
instant/reads/ByRun/sra/SRR/SRR168/SRR1686010/
SRR1686010.sra
$ fastq-dump --split-files SRR1686010.sra
```

Step 3: To download and prepare reference files

```
$ cd ../
# downloading human hg19 genome from Illumina
iGenomes
$ wgetftp://igenome:G3nom3s4u@ussd-ftp.illumina.com/
Homo_sapiens/UCSC/hg19/Homo_sapiens_UCSC_hg19.tar.
gz

# decompressing .gz files
$ tar -zxvf Homo_sapiens_UCSC_hg19.tar.gz
```

Step 4: To assemble super-reads

```
```

If your RNA-seq data are paired, you could use superreads.pl script to reconstruct the RNA-seq fragments from their end sequences, which we call super-reads. The superreads.pl script uses MaSuRCA genome assembler to identify pairs of reads that belong to the same super-read and extract the sequence containing the pair plus the sequence between them. Before running super-reads, install MaSuRCA. Input files are two paired-end *.fastq files, and output files are one super-reads *.fastq file (LongReads.fq.gz) and two notAssembled*.fastq files (SRR1686010_1.notAssembled.fq.gz and SRR1686010_2.notAssembled.fq.gz).

```
# create file named sr_config_example.txt that
contain below contents and put into the
<masurca_directory>.
*******************************************************
DATA
PE= pe 180 20 R1_001.fastq R2_001.fastq
JUMP= sh 3600 200 /FULL_PATH/short_1.fastq /FULL_
PATH/short_2.fastq
OTHER=/FULL_PATH/file.frg
END
PARAMETERS
GRAPH_KMER_SIZE=auto

USE_LINKING_MATES=1

LIMIT_JUMP_COVERAGE = 60

CA_PARAMETERS = ovlMerSize=30 cgwErrorRate=0.25
ovlMemory=4GB

NUM_THREADS= 64

JF_SIZE=100000000

DO_HOMOPOLYMER_TRIM=0
END
*******************************************************

$ cd SRR1686010
# copy superreads.pl scripts into SRR1686010
$ cp ../superreads.pl superreads.pl
# run superreads.pl to identify superreads
```

```
$ perl superreads.pl SRR1686010_1.fastq
SRR1686010_2.fastq <masurca_directory>
$ cp superreads.pl ../SRR1686013
$ cd ../SRR1686013
$ perl superreads.pl SRR1686013_1.fastq
SRR1686013_2.fastq <masurca_directory>
```

Step 5: To align assemble and non-assemble reads to the human reference sequence using TopHat 2

TopHat will be used to align super-reads and no assembled pair-end reads into the human genome and reference annotation. The GTF, genome index, and FASTQ files will be used as input files. When TopHat completes the analysis, accepted_hits.bam, align_summary.txt, deletions.bed, insertions.bed, junctions.bed, logs, prep_reads.info, and unmapped.bam files will be produced. The align_summary.txt contains summary of alignment. The accepted_hits.bam contains list of read alignment which will be used to assemble transcripts for each samples.

```
$ cd ../SRR1686010
# align super-reads and not Assembled pair-end
reads to genome and gene and transcript models
$ tophat -p 8 -G Homo_sapiens/UCSC/hg19/Annotation/
Genes/genes.gtfHomo_sapiens/UCSC/hg19/Sequence/
Bowtie2Index/genome SRR1686010_1.notAssembled.fq.gz
SRR1686010_2.notAssembled.fq.gz LongReads.fq.gz

$ cd ../SRR1686013
$ tophat -p 8 -G Homo_sapiens/UCSC/hg19/Annotation/
Genes/genes.gtfHomo_sapiens/UCSC/hg19/Sequence/
Bowtie2Index/genome SRR1686013_1.notAssembled.fq.gz
SRR1686013_2.notAssembled.fq.gz LongReads.fq.gz
```

Step 6: To assemble transcriptome by StringTie

StringTie assembles genes and transcripts (GTF) for each sample from read alignment files (BAM). The human gene and transcript models (genes.gtf) can be used as reference annotation to guide assembly. The SRR1686010.gtf and SRR1686013.gtf will be produced as output after finishing StringTie. The GTF files list all assembled genes and transcripts for each sample and it will be used as input for Cuffmerge.

```
$ cd ../SRR1686010
# runStringTie to assemble transcriptome
$ stringtietophat_out/accepted_hits.bam -o
SRR1686010.gtf -p 8 -G Homo_sapiens/UCSC/hg19/
Annotation/Genes/genes.gtf

$ cd ../SRR1686013
$ stringtietophat_out/accepted_hits.bam -o
SRR1686013.gtf -p 8 -G Homo_sapiens/UCSC/hg19/
Annotation/Genes/genes.gtf
```

Step 7: To merge two transcriptomes by Cuffmerge

When StringTie assembles the two transcriptomes separately, it will produce two different gene and transcript model files for each sample. Based on this, it is hard to compare expression between groups. Cuffmerge will assemble those transcript and gene models into a single comprehensive transcriptome. At first, you need to create a new text file which contains two GTF file addresses. Cuffmerge will then merge the two GTF files with the human reference GTF file and produce a single merged.gtf, which contains an assembly that merges all transcripts and genes in the two samples.

```
$ cd ../
# create a text file named assemble.txt that list
GTF files for each sample, Like:
****************************************************
SRR1686010/SRR1686010.gtf
SRR1686013/SRR1686013.gtf
****************************************************

# runcuffmerge to assemble a single GTF
$ cuffmerge -g Homo_sapiens/UCSC/hg19/Annotation/
Genes/genes.gtf -p 8 assemble.txt
```

Step 8: To identify differentially expressed genes and transcripts between decidual stromal cells and endometrial stromal fibroblasts by Cuffdiff

---

Cuffdiff will test the statistical significant transcripts and genes between groups. Two read alignment files (BAM) and one merged GTF will be used as input for cuffdiff. It will produce a number of output files that contain FPKM tracking files, count tracking files, read group tracking files, differential expression files, and run.info. The FPKM and count tracking files will generate FPKM and number of fragments of isoform, gene, cds, and primary transcripts in the two samples. The read group tracking files count fragments of isoform, gene, cds, and primary transcripts in two groups. The differential expression files list the statistical significant levels of isoform, gene, cds, primary transcript, promoter, and splicing between groups. Significant equal to *yes* depending on *p*-values after Benjamini–Hochberg correction for multiple tests is smaller than .05, which means those isoforms, genes, cds, promoters, and splicings have significant differential expression.

```
# identifying differentially expression genes and
transcripts
$ cuffdiff -o cuffdiff -p 8 merged.gtf SRR1686010/
tophat_out/accepted_hits.bam SRR1686013/tophat_out/
accepted_hits.bam
```

Note:

1. The parameter *p* means how many threads will be used in those commands. You can adjust the number following your computer resource.

2. $ means command for each step.

3. # means explains for each step.

More details follow in http://ccb.jhu.edu/software/tophat/manual.shtml; http://ccb.jhu.edu/software/stringtie/; and http://cufflinks.cbcb.umd.edu/manual.html.

5.4.2 Tutorial 2: BaseSpace® RNA Express Graphical User Interface

Illumina has developed BaseSpace®, a cloud-based genomics analysis workflow, which is integrated into the MiSeq, NextSeq, and HiSeq platforms. The cloud base platform eliminates the need for an on-site cluster and facilitates easy access to and sharing of data. During the sequencing run on an Illumina machine, the bcl files are automatically transferred to the users BaseSpace® account, where they are demultiplexed and converted into fastq files. For those users who require more in-depth command line base analyses, the bcl files can be simultaneously transferred to a local cluster. In addition, fastq files from previous runs and/or non-Illumina platforms can be imported into BaseSpace® for further analysis.

The graphics of BaseSpace® are modeled after the application icons made popular by Android and Apple operating systems. Analysis applications (apps) are available from both Illumina and third-party developers. Access to and storage in BaseSpace® is free; however, it does require registration. The use of the apps is either free or requires a nominal fee. Currently, BaseSpace® offers TopHat, Cufflinks, and RNA Express apps for RNA-seq analysis. Since we have already described the command lines for TopHat and Cufflinks, we will discuss the RNA Express GUI app in this section. The BaseSpace® RNA Express app combines the STAR aligner and DE-Seq analysis software, two commonly used workflows, into a single pipeline.

Log in and/or create your free BaseSpace® user account (https://basespace. illumina.com).

Step 1: *To create a project.* Click on the **Projects** icon and then the **New Projects** icon. Enter the name and description of your project and click **Create**.

Step 2: *To import data.* You can add samples (*.fastq files) to a project directly from an Illumina sequencing run or you can import files from a previous run. In our example, you will analyze the 4 *.fastq files representing the same RNA-seq data used for the Enhanced Tuxedo Suite Tutorial. Launch the SRA Import v0.0.3 app. Enter your project and the SRA# for the file to import (e.g., 1686013 and 1686010) and click **Continue**. These files should import within 30–60 min. Illumina will send you an e-mail when the files have been imported. Basespace will automatically filter and join the paired-end read files.

Step 3: *To launch the RNA Express app.* Once you have created your project and imported the *.fastq files, you are ready to run the **RNA Express** app. This app is currently limited to analysis of human, mouse, and rat reference genomes, but we are using RNA isolated from human, so we can proceed. While you have your project page open, click the **Launch** app icon. Select the **RNA Express** app. Under sample criteria, select the reference genome: **Homo sapiens/hg19**. **Check** the box for Stranded and Trim TruSeq Adapters. Under Control Group, select the control endometrial stromal fibroblasts files: **ES**. Click **Confirm**. Under Comparison Group, select the decidual stromal cells files: **DS**. Click **Confirm**. Select **Continue**. Your analysis will begin automatically. You will receive an e-mail notification when the analysis is complete.

Step 4: *To view the data analysis results.* Open your **Projects** page and select the **Analyses** link. Select the **RNA Express** link. A new page with the following types of information will be presented: Primary Analysis Information, Alignment Information, Read Counts, Differential Expression, Sample Correlation Matrix, Control vs. Comparison plot, and a Table listing the differentially expressed genes. The Control vs. Comparison plot and the Table are interactive, so you can select for the desired fold change and significance cutoffs. The data can be downloaded in both PDF and Excel formats for further analysis and figure presentation.

BIBLIOGRAPHY

Cheranova D, Gibson M, Chaudhary S, Zhang LQ, Heruth DP, Grigoryev DN, Ye SQ. RNA-seq analysis of transcriptomes in thrombin-treated and control human pulmonary microvascular endothelial cells. *J Vis Exp.* 2013; (72). pii: 4393. doi:10.3791/4393.

Cheranova D, Zhang LQ, Heruth D, Ye SQ. Chapter 6: Application of next-generation DNA sequencing in medical discovery. In *Bioinformatics: Genome Bioinformatics and Computational Biology.* 1st ed., pp. 123–136, Tuteja R (ed), Nova Science Publishers, Hauppauge, NY, 2012.

Finotello F, Di Camillo B. Measuring differential gene expression with RNA-seq: Challenges and strategies for data analysis. *Brief Funct Genomics.* September 18, 2014. pii: elu035. [Epub ahead of print] Review. PMID:25240000.

Kim D, Pertea G, Trapnell C, Pimentel H, Kelley R, Salzberg SL. TopHat2: Accurate alignment of transcriptomes in the presence of insertions, deletions and gene fusions. *Genome Biol.* 2013; 14:R36.

Korpelainen E, Tuimala J, Somervuo P, Huss M, Wong G (eds). *RNA-Seq Data Analysis: A Practical Approach.* Taylor & Francis Group, New York, 2015.

Mutz KO, Heilkenbrinker A, Lönne M, Walter JG, Stahl F. Transcriptome analysis using next-generation sequencing. *Curr Opin Biotechnol.* 2013; 24(1):22–30.

Rapaport F, Khanin R, Liang Y, Pirun M, Krek A, Zumbo P, Mason CE, Socci ND, Betel D. Comprehensive evaluation of differential gene expression analysis methods for RNA-seq data. *Genome Biol.* 2013; 14(9):R95.

Trapnell C, Williams BA, Pertea G, Mortazavi A, Kwan G, van Baren MJ, Salzberg SL, Wold BJ, Pachter L. Transcript assembly and quantification by RNA-seq reveals unannotated transcripts and isoform switching during cell differentiation. *Nat Biotechnol.* 2010; 28(5):511–515.

Wang Z, Gerstein M, Snyder M. RNA-seq: A revolutionary tool for transcriptomics. *Nat Rev Genet.* 2009; 10(1):57–63.

Zhou X, Ren L, Meng Q, Li Y, Yu Y, Yu J. The next-generation sequencing technology and application. *Protein Cell.* 2010; 1(6):520–536.

## CHAPTER 6

Microbiome-Seq Data Analysis

Daniel P. Heruth, Min Xiong, and Xun Jiang

CONTENTS

6.1 INTRODUCTION

Microbiome-sequencing (Microbiome-seq) is a technology that uses targeted, gene-specific next-generation sequencing (NGS) to determine both the diversity and abundance of all microbial cells, termed the microbiota, within a biological sample. Microbiome-seq involves sample collection and processing, innovative NGS technologies, and robust bioinformatics analyses. Microbiome-seq is often confused with metagenomic-seq, as the terms *microbiome* and *metagenome* are frequently used interchangeably; however, they describe distinct approaches to characterizing microbial communities. Microbiome-seq provides a profile of the microbial taxonomy within a sample, while metagenomic-seq reveals the composition of microbial genes within a sample. Although microbiome-seq and metagenomic-seq share common experimental and analytical strategies, in this chapter, we will focus on the analysis of microbiome-seq data.

A major advantage for microbiome-seq is that samples do not have to be cultured prior to analysis, thus allowing scientists the ability to rapidly characterize the phylogeny and taxonomy of microbial communities that in the past were difficult or impossible to study. For example, bacteria, typically the most numerous microorganisms in biological samples, are extremely difficult to culture, with estimates that less than 30% of bacteria collected from environmental samples can actually be cultured. Thus, advances in NGS and bioinformatics have facilitated a revolution in microbial ecology. The newly discovered diversity and variability of microbiota within and between biological samples are vast. To advance further the discovery and characterization of the global microbiota, several large projects, including the Earth Microbiome Project (www.earthmicrobiome.org), MetaHIT (www.metahit.eu), and the Human Microbiome Project (www.hmpdacc.org), have been established. In addition to coordinating and advancing efforts to characterize microbial communities from a wide array of environmental and animal samples, these projects have standardized the protocols for sample isolation and processing. This is a critical step in microbiome-seq to ensure that the diversity and variability between samples is authentic and not due to differences in the collection and handling of the samples. If a sample is not processed appropriately, the profile of the microbiota may not be representative of the original sample. For instance, if a stool sample is left at room temp and exposed to room air for even a short period of time, aerobic bacteria may continue to grow, while strict anaerobic bacteria will begin to die, thus skewing the taxonomic characterization of the sample. Therefore, we strongly encourage you to review the guidelines for sample isolation and processing prior to initiating a microbiome-seq project.

Microbiome-seq relies on the targeted sequencing of a single phylogenetic marker. The most commonly used marker is the ribosomal small subunit (SSU), termed 16S ribosomal RNA (rRNA) in archaea and bacteria, and 18S rRNA in eukaryotes. Additional phylogenetic markers are 5S rRNA, 23S rRNA, and the ribosomal internal transcribed spacer (ITS). ITS sequences are used for classifying fungal communities. These highly conserved ribosomal sequences share the same function in translation in all organisms, thus, providing an excellent phylogenetic marker. The ~1500 bp 16S rRNA gene contains nine different hypervariable regions flanked by evolutionarily conserved sequences. Universal primers complementary to the conserved regions ensure that the polymerase chain reaction (PCR) amplification of the DNA isolated from the experimental samples will generate amplicons of the desired variable region(s) representative for each type of bacterium present in the specimen. The resulting amplicons will contain the variable regions which will provide the genetic fingerprint used for taxonomic classification. The hypervariable regions between bacteria are frequently diverse enough to identify individual species. Primers targeting the variable V3 and V4 regions are most commonly used, although no region has been declared the best for phylogenetic analysis. We recommend reviewing the literature to determine which hypervariable regions are suggested for your specific biological samples.

The NGS platforms used for microbiome-seq are the same as those utilized for whole genome-seq and RNA-seq as described in Chapters 4 and 5, respectively. Roche 454 pyrosequencing (http://www.454.com) was the initial workhorse for microbiome-seq, however, due to advances in Illumina's (MiSeq, HiSeq; http://www.illumina.com) and Life Technologies' (Ion Torrent; http://www.lifetechnologies.com) platforms and chemistries, they are now commonly used in microbiome-seq. The advantage of Illumina's systems, in addition to lower costs and more coverage than 454 sequencing, is the ability to perform paired-end reads (MiSeq, 2 × 300; HiSeq 2500, 2 × 250) on PCR amplicons. However, since longer reads lead to more accurate taxonomic classifications, PacBio's RSII platform (http://www.pacificbiosciences.com/) may soon become the preferred platform for microbiome-seq. Dependent upon the NGS platform you use, there are several options available for the inclusion of sample identifying barcodes and heterogeneity spacers within the PCR amplicons, so review the latest sequencing protocols prior to initiating your experiment.

6.2 MICROBIOME-SEQ APPLICATIONS

Microorganisms constitute not only a large portion of the Earth's biomass, but they have also colonized eukaryotic organisms, including the gastrointestinal tract, oral cavity, skin, airway passages, and urogenital system. Table 6.1 lists several key representative applications of microbiome-seq.

TABLE 6.1   Microbiome-Seq Applications

#	Usages	Descriptions	References
1	Human gut microbiome	Difference in gut microbial communities	Yatsunenko et al. (2012)
		Nutrition, microbiome, immune system axis	Kau et al. (2011)
		Impact of diet on gut microbiota	De Filippo et al. (2010)
		Antibiotic perturbation	Dethlefsen & Relman (2011)
2	Human skin microbiome	Analysis of microbial communities from distinct skin sites	Grice et al. (2009)
3	Human nasal and oral microbiome	Comparison of microbiome between nasal and oral cavities in healthy humans	Bassis et al. (2014)
4	Human urinary tract microbiome	Urine microbiotas in adolescent males	Nelson et al. (2012)
5	Human placenta microbiome	Placental microbiome in 320 subjects	Aagaard et al. (2014)
6	Disease and microbiome	Crohn's disease	Eckburg and Relman (2007)
		Obesity	Turnbaugh et al. (2009)
		Colon cancer	Dejea et al. (2014)
7	Identification of new bacteria	Identification of mycobacterium in upper respiratory tract in healthy humans	Macovei et al. (2015)
8	Environmental classification	Deep-ocean thermal vent microbial communities	Reed et al. (2015)
		Root-associated micriobiome in rice	Edwards et al. (2015)
		Tallgrass prairie soil microbiome	Fierer et al. (2013)

6.3 DATA ANALYSIS OUTLINE

Microbiome-seq generates an enormous amount of data; a MiSeq (2×300) paired-end run produces 44–50 million reads passing filter, while a HiSeq 2500 (2×250) can generate more than 1.2 billion reads in a paired-end run. Therefore, an assortment of powerful statistical methods and computational pipelines are needed for the analysis of the microbiome-seq data. Several analysis pipelines for targeted-amplicon sequencing have been developed, including QIIME (www.qiime.org), QWRAP (https://github.

com/QWRAP/QWRAP), mothur (https://www.mothur.org), VAMPS (https://vamps.mbl.edu/index.php), and CloVR-16S (https://clovr.org/methods/clovr-16s/). In addition, there are numerous online resources, including Biostars (https://www.biostars.org) and Galaxy (https://www.galaxyproject.org), which provide both bioinformatics tutorials and discussion boards. The selection of an analysis pipeline will be dictated by the user's comfort with either Unix-based command line or graphical user interface platforms. Command line analysis workflows (QIIME and mothur) are the norm for microbiome-seq analysis; however, graphical user interface (GUI) software packages, such as Illumina's MiSeq Reporter Software and BaseSpace® (https://basespace.illumina.com), are growing in popularity. As NGS and microbiome-seq technologies are developed further, the rich resource of analysis pipelines will also continue to become both more powerful and user-friendly. The next challenge will be the development of software capable of performing large meta-analysis projects to capitalize fully on the ever increasing and diverse microbiome-seq data sets. The objective of this section is to provide a general outline to commonly encountered steps one faces on the path from raw microbiome-seq data to biological conclusions. For the ease of discussion, we will focus more specifically on Illumina sequencing technologies coupled with the QIIME analysis pipeline; however, the basic concepts are applicable to most microbiome-seq data analysis pipelines. QIIME is a collection of several third-party algorithms, so there are frequently numerous command options for each step in the data analysis. Figure 6.1 provides an example workflow for microbiome-seq.

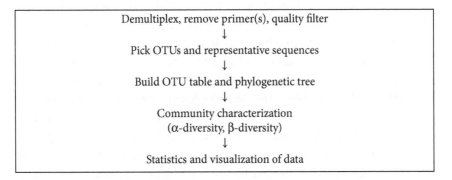

FIGURE 6.1 QIIME work flow for microbiome-seq data analysis. See text for brief description of each step.

Step 1: *Demultiplex, remove primer(s), quality filter.* High-throughput Illumina microbiome-seq allows multiple samples (100s to 1000s) to be analyzed in a single run. Samples are distinguished from one another by dual-indexing of the PCR amplicons with two unique 8 nt barcodes by adapter ligation during the PCR amplification steps of library preparation. In the case of Illumina sequencing, utilization of distinct barcodes facilitates adequate cluster discrimination. The addition of a heterogeneity spacer (0–7 nt) immediately downstream of the R1 barcode further enhances cluster discrimination throughout the sequencing run. The first steps in processing Illumina sequencing files are to convert the base call files (*.bcl) into *.fastq files and to demultiplex the samples. After paired-end sequencing, each read may be linked back to its original sample via its unique barcode. Illumina's bcl2fastq2 Conversion Software v2.17.1.14 can demultiplex multiplexed samples during the step converting *.bcl files into *.fastq.gz files (compressed FASTQ files). The MiSeq Reporter and BaseSpace® software automatically demultiplex and convert *.bcl files to *.fastq files. QIIME (Quantitative Insights into Microbial Ecology) analysis starts with the *.fastq files and a user-generated mapping file. The mapping file is a tab-delimited text doc containing the sample name, barcodes, and sequencing primer. Like the bcl2fastq2 Conversion Software, QIIME can also demultiplex fastq files. QIIME uses the mapping file to assign the *.fastq sequences to their appropriate sample. QIIME removes both the barcodes and primer sequences from the *.fastq files, so the remaining sequence is specific to the 16S rRNA hypervariable region. The final steps in preprocessing of sequence data are to determine the quality (filter) of the sequence, convert the *.fastq files to fasta formatted files (*.fsn), and generate the reverse complement of the R2 sequences. Quality filtering removes sequences that contain low quality (e.g., $Q < 30$) and more than one ambiguous call (N). QIIME combines the demultiplexing, primer removal, quality filtering, reverse complementation, and conversion to fasta (*.fsn) files with a single QIIME command (split_libraries_fastq.py). A positive of Illumina sequencing is that overlapping paired-end sequences can be merged to generate larger sequences for analysis. Several programs, such as PANDAseq (https://github.com/neufeld/pandaseq) and PEAR (http://www.exelixis-lab.org/web/software/pear), are available for joining paired end reads. Finally,

chimeric sequences need to be removed. Chimeras are PCR artifacts generated during library amplification which result in an
overestimation of community diversity. Chimeras can be removed
during the processing step or following operational taxonomic
unit (OTU) picking in step 2.

Step 2: *Pick OTUs and representative sequences.* Once the sequences
have been processed, the 16S rRNA amplicon sequences are assigned
into OTUs, which are based upon their similarity to other sequences
in the sample. This step, called *OTU picking*, clusters the sequences
together into identity thresholds, typically 97% sequence homology,
which is assumed to represent a common species. There are three
approaches to OTU picking; *de novo,* closed-reference, and open-
reference. *De novo* OTU picking (pick_de_novo_otus_py) clusters
sequences against each other with no comparison to an external reference database. Closed-reference OTU picking (pick_closed_refer-
ence_otus.py) clusters the sequences to a reference database and any
non-matching sequences are discarded. Open-reference OTU picking (pick_open_reference_otus.py) clusters sequences to a reference
database and any non-matching sequences are then clustered using
the *de novo* approach. Open-reference OTU picking is the most
commonly used method, although we recommend reviewing the
QIIME OTU tutorial (http://qiime.org/tutorials/otu_picking.html)
prior to sequence analysis. Each OTU will contain hundreds of clustered sequences, so a representative sequence for each OTU will be
selected to speed up the downstream analyses.

Step 3: *Build OTU table and phylogenetic tree.* Each representative
OTU is now assigned taxonomically to a known organism using a
reference database, such as GreenGenes (http://greengenes.lbl.gov/),
Ribosomal Database Project (RDP) (http://rdp.cme.msu.edu/), and
Silva (http://www.arb-silva.de/). The sequences are then aligned further against the reference database and a phylogenetic tree is inferred
from the multiple sequence alignment. The final step is to build the
OTU table which presents the taxonomic summary, including relative abundance of the sequencing results at the kingdom, phylum,
class, order, family, genus, and species levels. Based upon the hypervariable region(s) which were sequenced and the specific characteristics of individual bacteria, it may not always be possible to identify
sequences to the genus and/or species levels.

Step 4: *Community classification.* The OTU table, phylogenetic tree, and mapping file are used to classify the diversity of organisms within and between the sequenced samples. α-diversity is defined as the diversity of organisms within a sample, while β-diversity is the differences in diversity between samples.

Step 5: *Statistics and visualization of data.* To facilitate the dissemination of a microbiome-seq experiment, QIIME generates statistics at each step of the analysis workflow (OTU table, phylogenetic tree, α-diversity, and β-diversity), as well as visualization tools.

6.4 STEP-BY-STEP TUTORIAL FOR MICROBIOME-SEQ DATA ANALYSIS

In this section, we will demonstrate step-by-step tutorials on two distinct microbiome-seq data analysis workflows. First, we will present a QIIME command line pipeline utilizing publically available MiSeq data, followed by the introduction of 16S Metagenomics v1.0, a GUI workflow available on Illumina's BaseSpace®.

6.4.1 Tutorial 1: QIIME Command Line Pipeline

Here, we present the QIIME workflow for in-depth analysis of RNA-seq data. Command line-based pipelines, like QIIME, typically require a local cluster for both the analysis and storage of data; however, QIIME also provides Windows virtual box (http://qiime.org/install/virtual_box.html) and MacQIIME (http://www.wernerlab.org/) for your consideration. Here, we provide a sample tutorial with MiSeq (V4; 2 × 250) data you can use for practice.

Step 1: To download the required programs

----

a. QIIME (QIIME.org).

b. USEARCH (http://www.drive5.com/usearch/). Rename the 32 bit binary file to usearch61.

c. PEAR (http://www.exelixis-lab.org/web/software/pear).

d. Python script (https://github.com/ThachRocky/QIIME_FASTQ_TO_FASTA). Specialized script coded for this analysis.

e. BioPython script (http://biopython.org/)

f. GreenGenes reference sequences. (ftp://greengenes.microbio.me/
greengenes_release/gg_13_5/gg_13_8_otus.tar.gz)

g. FigTree Viewer (http://tree.bio.ed.ac.uk/software/figtree/)

h. MiSeq sequence files:
```
$wget ftp://ftp-trace.ncbi.nlm.nih.gov/sra/
sra-instant/reads/ByRun/sra/SRR/SRR651/
SRR651334/SRR651334.sra
$wget ftp://ftp-trace.ncbi.nlm.nih.gov/sra/
sra-instant/reads/ByRun/sra/SRR/SRR104/
SRR1047080/SRR1047080.sra
```

Step 2: To create fastq files for each sample you downloaded

We have selected two samples from a larger project (SRP001634) to study the metagenome from infant gut. These two datasets represent the micriobiome of an individual infant at 2 and 3 weeks of life. You can read more about the study at http://www.ncbi.nlm.nih.gov/bioproject/PRJNA63661. The first step is to create *.fastq files from the *.sra files you downloaded. These are 2 × 175 bp reads.

```
$fastq-dump --split-3 SRR651334
$fastq-dump --split-3 SRR1047080
```

Note: Two output *.fastq files will be generated for each command which represent the forward and reverse sequencing reads. For example, the files for SRR651334 will be: SRR651334_1.fastq and reverse SRR651334_2.fastq

Step 3: To join paired ends

The next step is to join the two paired-end sequencing *.fastq files generated in step 1 using the PEAR software. This step includes quality filtering and generating the complement of the reverse sequence.

```
$pear -f SRR651334_1.fastq -r SRR651334_2.fastq -n
250 -q 38 -o SRR651334
$pear -f SRR1047080_1.fastq -r SRR1047080_2.fastq
-n 250 -q 38 -o SRR1047080
```

Note: The anatomy of this command:

–n: Specify the minimum length of assembled sequence. We just want to take the successfully joined sequence that is why we set –n = 250 bp.

–q: Specify the quality score threshold for trimming the low quality part of the read. In this case, for the maximum size of 350 bp, we recommend to use the –q = 38

–f: forward reads

–r: reverse reads

Note: The output files will be SRR651334.assembled.fastq and SRR1047080. assembled.fastq.

Step 4: Clean and convert joined *fastq files to fasta

Utilize the specialized python script to convert files to *.fasta format, which is necessary for downstream QIIME analysis.

```
# Executable format:
#python Clean_Convert_Fastq_to_Fasta.py <fastq_
file> <new_name.fasta>

$python Clean_Convert_Fastq_to_Fasta.py SRR651334.
assembled.fastq SRR651334.fasta
$python Clean_Convert_Fastq_to_Fasta.py SRR1047080.
assembled.fastq SRR1047080.fasta
```

Note: The output files will be SRR651334.fasta and SRR1047080.fasta.

Step 5: To create mapping file

Your must now create a mapping file which contains the following information: sample ID, barcode sequence, linker primer sequence, input file name, and description. If you are analyzing original data, you must create your mapping file when you are demultiplexing your sequences. However, since you are analyzing SRA-deposited sequences, this is the step to create the mapping file. Since these sequences have already been demultiplexed, you do not need to enter the barcode and primer sequences; however, the heading must be present in the text file. The sample IDs and input file names are mandatory, and the descriptions are recommended. This is a tricky part because each heading is required and must be tab-separated. If the downstream step does not work, recheck your mapping file format.

```
#SampleID   BarcodeSequence LinkerPrimerSequence InputFileName Description
SRR651334                                        SRR651334.fasta    week2
SRR1047080                                       SRR1047080.fasta   week3
```

Note: Your format will look like: "SRR651334\t<blank>\t<blank>\t<SRR651334.fasta>\t<week3>\n" t = tab and n = end of the line. Save your mapping file as <mapping_file.txt>.

Step 6: To add QIIME labels

The first step is to create a new folder containing both SRR651334.fasta and SRR1047080.fasta files, followed by the QIIME command to combine the files and add the information listed in the mapping_txt file.

```
$mkdir merged_reads
$cp SRR651334.fasta merged_reads/
$cp SRR1047080.fasta merged_reads/
```

Add qiime label

```
$add_qiime_labels.py -i merged_reads/ - m mapping_
file.txt -c InputFileName -n 1 -o Combined_fasta/
```

Note: The output folder will be named combined_fasta. You can check the contents with the following command:

```
$ls -l Combined_fasta/
```

A single fasta file, combined_seqs.fna, will be present.

Step 7: To check and remove chimeric sequences

The combined_seqs.fna file should be screened to remove chimeras. QIIME currently includes a taxonomy-assignment-based approach, blast_fragments, for identifying chimeric sequences. The chimera running code requires the rep_set_aligned reference. We use the GreenGenes reference library gg_13_8_otus/rep_set_aligned/99_otus.fasta. We recommend using 99% homology rather than 97%, because the fasta files reported with 97% homology will contain dashes in place of uncalled nucleotides.

```
$identify_chimeric_seqs.py
-i Combined_fasta/combined_seqs.fna
-r /data/reference/Qiime_data_files/gg_13_8_otus/
rep_set_aligned/99_otus.fasta
-m usearch61
-o Combined_fasta/usearch_checked_chimeras/

$filter_fasta.py
-f Combined_fasta/combined_seqs.fna
-o Combined_fasta/seqs_chimeras_filtered.fna
-s Combined_fasta/usearch_checked_chimeras/chimeras.txt
-n
```

Note: Each of the two commands listed above should be entered as a single line. There should be a single space between the command and the next parameter. When the first command is completed, you may run the second command. Check the output with the following command:

```
ls -l Combined_fasta/usearch_checked_chimeras/
```

The key output file <Combined_fasta/seqs_chimeras_filtered.fna> will be used in the next step.

Step 8: Pick OTUs

Now you are finally ready to begin the taxonomic classification of your sequence data by picking the OTUs. From now on, we will follow the format of step 7, where the commands are written in a single line.

```
# Executable syntax pick_otus.py

$pick_otus.py
-m usearch61
-i Combined_fasta/seqs_chimeras_filtered.fna
-o Combined_fasta/picked_otus_default/
```

Note: To check the ouput, use the command:

```
$ls -l Combined_fasta/picked_otus_default/
```

Note: The <Combined_fasta/picked_otus_default/seqs_chimeras_filtered_ otus.txt> file will be used in both steps 9 and 11.

Step 9: To pick representation set

This step picks a representative sequence set, one sequence from each OTU. This step will generate a *de novo* fasta (fna) file for each representation set of OTUs, named default_rep.fna.

```
# Executable syntax pick_rep_set.py

$pick_rep_set.py
-i Combined_fasta/picked_otus_default/seqs_chimeras_
filtered_otus.txt
-f Combined_fasta/seqs.fna
-o Combined_fasta/default_rep.fna
```

Step 10: Assign Taxonomy

This step requires that you know the path of the GreenGenes reference data set. Given a set of sequences, the command assign_taxonomy.py attempts to assign the taxonomy of each sequence. The output of this step is an observation metadata mapping file of input sequence identifiers (1st column of output file) to taxonomy (2nd column) and quality score (3rd column). There may be method-specific information in subsequent columns. The standard practice utilizes the 97% threshold to determine homology.

```
Executable syntax assign_taxonomy.py

$assign_taxonomy.py
-i Combined_fasta/default_rep.fna
-r gg_13_8_otus/rep_set/97_otus.fasta
-t gg_13_8_otus/taxonomy/97_otu_taxonomy.txt
-o Combined_fasta/taxonomy_results/
```

Note: Check the output files

```
$ls -l Combined_fasta/taxonomy_results/
```

The < Combined _ fasta/taxonomy _ results/default _ rep _ tax _ assignments.txt > file will be used in step 11.

Step 11: Make OTUS table

The script make_otu_table.py tabulates the number of times an OTU is found in each sample and adds the taxonomic predictions for each OTU in the last column if a taxonomy file is supplied. The –i text file was generated in step 8 and the –t file was generated in step 10.

```
Executable syntax make_otu_table.py
$make_otu_table.py
-i Combined_fasta/picked_otus_default/seqs_chimeras_
filtered_otus.txt
-t Combined_fasta/taxonomy_results/default_rep_tax_
assignments.txt
-o Combined_fasta/otu_table.biom
```

Step 12: To summarize results

The summarize_taxa.py script provides summary information of the representation of taxonomic groups within each sample. It takes an OTU table (Combined_fasta/otu_table.biom) that contains taxonomic information as input. The taxonomic level for which the summary information is provided is designated with the –L option. The meaning of this level will depend on the format of the taxon strings that are returned from the taxonomy assignment step. The taxonomy strings that are most useful are those that standardize the taxonomic level with the depth in the taxonomic strings. For instance, for the RDP classifier taxonomy, Level 1 = Kingdom (e.g., Bacteria), 2 = Phylum (e.g., Firmicutes), 3 = Class (e.g., Clostridia), 4 = Order (e.g., Clostridiales), 5 = Family (e.g., Clostridiaceae), and 6 = Genus (e.g., Clostridium).

```
Executable syntax summarize_taxa.py

$summarize_taxa.py
-i Combined_fasta/otu_table.biom
-o Combined_fasta/taxonomy_summaries/

$ls -l Combined_fasta/taxonomy_summaries/
```

Step 13: To generate phylogenetic trees

To test the evolutionary distance between the OTUs, you can build a phylogenetic tree. This is a 3-step process that will take about 30 min to run. The three steps are to align the sequences to a reference database, quality filter the alignment, and generate the phylogenetic tree. There are several phylogentic tree viewing softwares available, and we recommend FigTree. It is very easy to install and use. You can use the $ls –l command to check the output file. The output file from each step will be used in the subsequent step.

```
Executable syntax align_seqs.py

$align_seqs.py
-i Combined_fasta/default_rep.fna
-t gg_13_8_otus/rep_set_aligned/97_otus.fasta
-o Combined_fasta/alignment/

Executable syntax filter_alignment.py
```

```
$filter_alignment.py
-i Combined_fasta/alignment/default_rep_
aligned.fasta
-o Combined_fasta/alignment/
```

```
Executable syntax make_phylogeny.py
```

```
$make_phylogeny.py
-i Combined_fasta/alignment/default_rep_aligned_
pfiltered.fasta
-o Combined_fasta/rep_set_tree.tre
```

```
Open the rep_set_tree.tre file in FigTree to view
the phylogenetic tree.
```

Step 14: To calculate alpha diversity

The QIIME script for calculating α-diversity in samples is called alpha_diversity.py. Remember, α-diversity is defined as the diversity of organisms within a sample.

```
# 1. Executable syntax multiple_rarefactions.py
$multiple_rarefactions.py
-i Combined_fasta/otu_table.biom
-m 100 -x 1000 -s 20 -n 10
-o Combined_fasta/rare_1000/
```

To check the file: $ls –l Combined_fasta/rare_1000/

```
# 2. Perform Calculate Alpha Diversity
$alpha_diversity.py
-i Combined_fasta/rare_1000/
-o Combined_fasta/alpha_rare/
-t Combined_fasta/rep_set_tree.tre
-m observed_species, chao1,PD_whole_tree, shannon
```

```
# 3. Summarize the Alpha Diversity Data
$collate_alpha.py
-i Combined_fasta/alpha_rare/
-o Combined_fasta/alpha_collated/
```

The results in alpha_collated are presented in tab-delimited text files.

Step 15: To calculate beta diversity

β-diversity is the differences in diversity between samples. You can perform weighted or unweighted unifrac analysis. We demonstrated weighted unifrac in this tutorial.

```
# Executable syntax beta_diversity.py
$beta_diversity.py
-i Combined_fasta/otu_table.biom
-m weighted_unifrac
-o Combined_fasta/beta_div/
-t Combined_fasta/rep_set_tree.tre
```

The results in beta_div are presented in tab-delimited text file table.

6.4.2 Tutorial 2: BaseSpace® 16S Metagenomics v1.0 Graphical User Interface

As described in Chapter 5, Illumina has developed BaseSpace®, a cloud-based genomics analysis workflow, which is integrated into the MiSeq, NextSeq, and HiSeq platforms. The cloud-based platform eliminates the need for an on-site cluster and facilitates easy access to and sharing of data. During the sequencing run on an Illumina machine, the *.bcl files are automatically transferred to the users BaseSpace® account, where they are demultiplexed and converted into *.fastq files. For those users who require more in-depth command line base analyses, the *.bcl files can be simultaneously transferred to a local cluster. In addition, *.fastq files from previous runs and/or non-Illumina platforms can be imported into BaseSpace® for further analysis. Currently, BaseSpace® offers the following apps for microbiome-seq analysis: 16S Metagenomics v1.0 and Kraken Metagenomics. We will discuss the 16S Metagenomics v1.0 GUI app in this section. The 16S Metagenomics v1.0 app utilizes the RDP classifier (https://rdp.cme.msu.edu/classifier/classifier.jsp) and an Illumina-curated version of the GreenGenes database to taxonomically classify 16S rRNA amplicon reads. The home page or dashboard for your personalized BaseSpace® account provides access to important notifications from Illumina, along with your runs, projects, and analyses.

Log in and/or create your free BaseSpace® user account (https://basespace.illumina.com).

Step 1: *To create a project.* Click on the **Projects** icon and then the **New Project** icon. Enter the name and description of your project and click **Create.**

Step 2: *To import data.* You can add samples (*fastq files) to a project directly from an Illumina sequencing run or you can import files from a previous run. In our example, you will analyze the same MiSeq *fastq files you used above in step 2 in the Qiime tutorial. Import these files, one at a time, by launching the SRA Import v0.0.3 app. Enter your project and the SRA# (651334 and 1047080), click **Continue.** These files should import within 30 min. Illumina will send you an e-mail when the files have been imported. BaseSpace® will automatically filter and join the paired-end read files.

Step 3: *To launch the 16S Metagenomics v1.0 app.* Once you have created your project and imported the sequence files, you are ready to run the **16S Metagenomics v1.0** app. While you have your project page open, click the **Launch app** icon. Select the **16S Metagenomics v1.0** app. Click **Select Samples**, select the files you wish to analyze. Click **Confirm.** Click **Continue.** Your analysis will begin automatically. You will receive an e-mail notification when the analysis is complete. Analysis of these two files will take approximately 30 min.

Step 4: *To view the data analysis results.* Open your **Projects** page and select the **Analyses** link. Select the **16S Metagenomics v1.0** link. A new page with the following types of information will be presented for both samples individually, along with an aggregate summary. The types of data presented are Sample Information, Classification Statistics, Sunburst Classification Chart, and the Top 20 Classification Results by Taxonomic Level. The data can be downloaded in both *.pdf and Excel formats for further analysis and figure presentation.

BIBLIOGRAPHY

Aagaard K, Ma J, Antony KM, Ganu R, Petrosino J et al. The placenta harbors a unique microbiome. *Sci Transl Med.*, 2014; 6(237), 237ra265. doi:10.1126/scitranslmed.3008599.

Bassis CM, Tang AL, Young VB, Pynnonen MA. The nasal cavity microbiota of healthy adults. *Microbiome*, 2014; 2:27. doi:10.1186/2049-2618-2-27.

Caporaso JG, Kuczynski J, Stombaugh J, Bittinger K, Bushman FD et al. QIIME allows analysis of high throughput community sequencing data. *Nat. Methods*, 2010; 7(5):335–336.

Caporaso JG, Lauber CL, Walters WA, Berg-Lyons D, Huntley J et al. Ultra-high-throughput microbial community analysis on the Illumina HiSeq and MiSeq platforms. *ISME J.*, 2012; 6(8):1621–1624.

Consortium HMP. Structure, function and diversity of the healthy human microbiome. *Nature*, 2012; 486:207–214.

De Filippo C, Cavalieri D, Di Paola M, Ramazzotti M, Poullet JB et al. Impact of diet in shaping gut microbiota revealed by a comparative study in children from Europe and rural Africa. *Proc Natl Acad Sci U S A*, 2010; 107(33):14691–14696.

Dejea CM, Wick EC, Hechenbleikner EM, White JR, Mark Welch JL et al. (2014). Microbiota organization is a distinct feature of proximal colorectal cancers. *Proc Natl Acad Sci U S A*, 2014; 111(51):18321–18326.

Dethlefsen L and Relman DA. Incomplete recovery and individualized responses of the human distal gut microbiota to repeated antibiotic perturbation. *Proc Natl Acad Sci U S A*, 2011; 108 Suppl 1:4554–4561.

Eckburg PB and Relman DA. The role of microbes in Crohn's disease. *Clin Infect Dis.*, 2007; 44(2):256–262.

Edwards J, Johnson C, Santos-Medellin C, Lurie E, Podishetty NK et al. Structure, variation, and assembly of the root-associated microbiomes of rice. *Proc Natl Acad Sci U S A*, 2015; 112(8):E911–920.

Fadrosh DW, Ma B, Gajer P, Sengamalay N, Ott S et al. An improved dual-indexing approach for multiplexed 16S rRNA gene sequencing on the Illumina MiSeq platform. *Microbiome*, 2014; 2(1):6. doi:10.1186/2049-2618-2-6.

Fierer N, Ladau J, Clemente JC, Leff JW, Owens SM et al. Reconstructing the microbial diversity and function of pre-agricultural tallgrass prairie soils in the United States. *Science*, 2013; 342(6158):621–624.

Gonzalez A, Knight R. Advancing analytical algorithms and pipelines for billions of microbial sequences. *Curr. Opin. Biotechnol.*, 2012; 23(1):64–71.

Grice EA, Kong HH, Conlan S, Deming CB, Davis J et al. Topographical and temporal diversity of the human skin microbiome. *Science*, 2009; 324(5931): 1190–1192.

Grice EA and Segre JA. The human microbiome: Our second genome. *Annu. Rev. Genomics Hum. Genet.*, 2012; 13:151–170.

Kau AL, Ahern PP, Griffin NW, Goodman AL, Gordon JI. Human nutrition, the gut microbiome and the immune system. *Nature*, 2011; 474(7351): 327–336.

Kozich JJ, Westcott SL, Baxter NT, Highlander SK, Schloss PD. Development of a dual-index sequencing strategy and curation pipeline for analyzing amplicon sequence data on the MiSeq Illumina sequencing platform. *Appl. Environ. Microbiol.*, 2013; 79(17):5112–5120.

Kuczynski J, Liu Z, Lozupone C, McDonald D, Fierer N et al. Microbial community resemblance methods differ in their ability to detect biologically relevant patterns. *Nature Methods*, 2010; 7(10):813–819.

Kuczynski J, Lauber CL, Walters WA, Parfrey LW, Clemente JC et al. Experimental and analytical tools for studying the human microbiome. *Nat. Rev. Genet.*, 2012; 13(1):47–58.

Kumar R, Eipers P, Little R, Crowley M, Crossman D et al. Getting started with microbiome analysis: Sample acquisition to bioinformatics. *Curr. Protocol. Hum. Genet.*, 2014; 82:18.8.1–18.8.29.

Macovei L, McCafferty J, Chen T, Teles F, Hasturk H et al. (2015). The hidden 'mycobacteriome' of the human healthy oral cavity and upper respiratory tract. *J Oral Microbiol.*, 2015; 7:26094. doi:10.3402/jom.v7.26094.

Navas-Molina JA, Peralta-Sánchez JM, González A, McMurdie PJ, Vázquez-Baeza Y et al. Advancing our understanding of the human microbiome using QIIME. *Methods Enzymol.* 2013; 531:371–444.

Schloss PD, Westcott SL, Ryabin T, Hall JR, Hartmann M et al. Introducing mothur: Open-source, platform-independent, community supported software for describing and comparing microbial communities. *Appl. Environ. Microbiol.* 2009; 75(23): 7537–7541.

Schwabe RF and Jobin C. The microbiome and cancer. *Nat. Rev. Can.*, 2013; 13: 800–812.

CHAPTER 7

miRNA-Seq Data Analysis

Daniel P. Heruth, Min Xiong, and Guang-Liang Bi

CONTENTS

7.1 INTRODUCTION

miRNA-sequencing (miRNA-seq) uses next-generation sequencing (NGS) technology to determine the identity and abundance of microRNA (miRNA) in biological samples. Originally discovered in nematodes, miRNAs are an endogenous class of small, non-coding RNA molecules that regulate critical cellular functions, including growth, development, apoptosis, and innate and adaptive immune responses. miRNAs negatively regulate gene expression by using partial complementary base pairing to target sequences in the 3′-untranslated region, and recently reported 5′-untranslated region, of messenger RNAs (mRNAs) to alter protein synthesis through either the degradation or translational inhibition of target mRNAs. miRNAs are synthesized from larger primary transcripts (pri-miRNAs), which, like mRNA, contain a 5′ cap and a 3′ poly-adenosine tail. The pri-miRNAs fold into hairpin structures that are subsequently cleaved in the nucleus by Drosha, an RNase III enzyme, into precursor miRNA (pre-miRNA) that are approximately 70 nucleotides in length and folded into a hairpin.

The pre-miRNA is transported to the cytoplasm where the hairpin structure is processed further by the RNase III enzyme Dicer to release the hairpin loop from the mature, double-stranded miRNA molecules. Mature miRNAs are approximately 22 nucleotide duplexes consisting of the mature guide miRNA, termed 5p, and the complementary star (*), termed 3p, miRNA. In vertebrates, the single-stranded guide miRNA is assembled into the RNA-induced silencing complex (RISC), which is guided to its mRNA target by the miRNA. The imperfect miRNA-mRNA base pairing destabilizes the mRNA transcript leading to decreased translation and/or stability. More than 1800 miRNAs have been identified in the human transcriptome (http://www.mirbase.org) with each miRNA predicted to regulate 5–10 different mRNAs. In addition, a single mRNA may be regulated by multiple miRNAs. Thus, miRNAs have the potential to significantly alter numerous gene expression networks.

Prior to the technological advances in NGS, microarrays and quantitative real-time polymerase chain reaction (qPCR) were the major platforms for the detection of miRNA in biological samples. Although these platforms remain as powerful tools for determining miRNA expression profiles, miRNA-seq is rapidly becoming the methodology of choice to simultaneously detect known miRNAs and discover novel miRNAs.

The NGS platforms used for miRNA-seq are the same as those utilized for whole-genome-seq and RNA-seq as described in Chapters 4 and 5, respectively. Illumina (MiSeq, HiSeq; http://www.illumina.com) and Life Technologies (Ion Torrent; http://www.lifetechnologies.com) continue to lead the field in developing the platforms and chemistries required for miRNA-seq. To prepare the miRNA for sequencing, 5′ and 3′ adapters are ligated onto the single-stranded miRNA in preparation for qPCR amplification to generate indexed miRNA libraries. The libraries are pooled, purified, and then subjected to high-throughput single-read (1 × 50 bp) sequencing. In addition to miRNA analyses, these methodologies also provide sequence information for additional small RNA molecules, including short-interfering RNA (siRNA) and piwi-interacting RNA (piRNA).

7.2 miRNA-SEQ APPLICATIONS

Regulation of miRNA expression is controlled at both cell- and tissue-specific levels. Thus, elucidating differential miRNA expression profiles could provide critical insights into complex biological processes, including global gene regulation and development. Several diseases have been

TABLE 7.1   miRNA-Seq Applications

#	Usages	Descriptions	References
1	Development	Animal development	Wienholds and Plasterk (2005)
		Lymphopoiesis	Kuchen et al. (2010)
		Cardiovascular system development	Liu and Olson (2010)
		Brain development	Somel et al. (2011)
2	Disease	Huntington's disease	Marti et al. (2010)
		Bladder cancer	Han et al. (2011)
		Kawasaki disease	Shimizu et al. (2013)
		Lung cancer	Ma et al. (2014)
3	Biomarkers	Tuberculosis	Zhang et al. (2014)
		Type 2 diabetes	Higuchi et al. (2015)
		Epilepsy	Wang et al. (2015)
4	Agriculture	Regulatory networks in apple	Xia et al. (2012)
		Leaf senescence in rice	Xu et al. (2014)
		Postpartum dairy cattle	Fatima et al. (2014)
5	Evolution	Zebrafish miRNome	Desvignes (2014)
		Genetic variability across species	Zorc et al. (2015)

associated with abnormal miRNA expression, including arthritis, cancer, heart disease, immunological disorders, and neurological diseases. As such, miRNAs have also been identified as promising biomarkers for disease. Table 7.1 lists several key representative applications of miRNA-seq.

7.3 miRNA-SEQ DATA ANALYSIS OUTLINE

The capacity of high-throughput, parallel sequencing afforded by the short, single-reads (1 × 50) utilized in miRNA-seq is a technological double-edged sword. One edge provides the advantages of highly multiplexed samples coupled with a low number of reads required for significant sequencing depth. The other edge presents the challenges of determining miRNA expression profiles in 100s of samples simultaneously, including the ability to distinguish accurately between short, highly conserved sequences, as well as the capability to distinguish mature and primary transcripts from degradation products. To address these challenges, numerous analysis pipelines have been developed, including miRDeep2 (www.mdc-berlin.de/8551903/en/), CAP-miRSeq (http://bioinformaticstools.mayo.edu/research/cap-mirseq/), miRNAkey (http://ibis.tau.ac.il/miRNAkey/), small RNA workbench (http://genboree.org), and miRanalyzer (http://bioinfo5.ugr.es/miRanalyzer/miRanalyzer.php). The list of available miRNA-seq analysis software packages is vast and continues to grow rapidly; thus, *it is not possible to cover all the approaches to analyzing miRNA-seq data. The objective of this section is to provide a general outline to commonly encountered steps and questions one faces on the path from raw miRNA-seq data to biological conclusions.* Figure 7.1 provides an example workflow for miRNA-seq.

Step 1: *Quality assessment and pre-processing.* High-throughput Illumina and Life Technologies miRNA-seq allow multiple samples (10s to 100s) to be analyzed in a single run. Samples are distinguished from one another by single-indexing of the PCR amplicons with unique barcodes by adapter ligation during the PCR amplification steps of library preparation. The first step in processing the sequencing files is to convert the base call files (*.bcl) into *.fastq files and to demultiplex the samples. After single-end sequencing, each read may be linked back to its original sample via its unique barcode. Illumina's bcl2fastq2 Conversion Software v2.17.1.14 can demultiplex multiplexed samples during the step converting *.bcl files into *.fastq.gz files (compressed FASTQ files). Life Technologies' Torrent Suite Software (v3.4) generates unmapped BAM files that can be converted into *.fastq files with the SamToFastq tool that is part of the Picard package. The fastq files (sequencing reads) are first quality-checked to remove low-quality bases from the 3' end and then processed further by trimming the PCR amplification adapters. The reads are quality filtered one more time to remove sequences that are <17 bases.

Step 2: *Alignment.* To identify both known and novel miRNAs, as well as to determine differential gene expression profiles, the reads

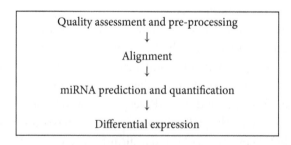

FIGURE 7.1   miRNA-seq data analysis pipeline. See text for a brief description of each step.

must first be aligned to the appropriate reference genome (i.e., human, mouse, and rat) and to a miRNA database, such as miRBase (http://www.mirbase.org/). The reads which map to multiple positions within a genome and/or map to known small RNA coordinates (e.g., snoRNA rRNA, tRNA), along with any reads that do not map to the reference genome, are discarded.

Step 3: *miRNA prediction and quantification.* The reads are evaluated for miRNAs which map to known miRNA gene coordinates and for novel sequences which possess characteristics of miRNA (e.g., energetic stability and secondary structure prediction). In addition, the read distribution of sequences aligned in step 2 (5′ end, hairpin structure, loop, 3′ end) is analyzed to distinguish between pre-miRNA and mature miRNA. Typically, a confidence score is assigned to each miRNA detected to facilitate further evaluation of the sequence data. Finally, the number of reads per miRNA is counted and then normalized to an RPKM expression index (reads per kilobase per million mapped reads) to allow comparison between samples and across experiments.

Step 4: *Differential expression.* NGS technologies, including miRNA-seq, provide digital gene expression data that can be used to determine differential expression profiles between two biological conditions. There are several software packages, such as edgeR (www.bioconductor .org), that use differential signal analyses to statistically predict gene expression profiles between samples. These data can be processed further for biological interpretation including gene ontology and pathway analysis.

7.4 STEP-BY-STEP TUTORIAL ON miRNA-SEQ DATA ANALYSIS

In this section, we will demonstrate step-by-step tutorials on two distinct miRNA-seq data analysis workflows. First, we will present the miRDeep2 command line workflow, followed by a tutorial on the small RNA workbench, a publically available GUI workflow. We will utilize the same publically available miRNA-seq data for both tutorials.

7.4.1 Tutorial 1: miRDeep2 Command Line Pipeline

miRDeep2 (www.mdc-berlin.de/8551903/en/) is a proven analysis workflow that couples the identification of both known and novel miRNAs with the determination of expression profiles across multiple samples.

The miRDeep2 algorithm, an enhanced version of miRDeep, utilizes a probabilistic model to analyze the structural features of small RNAs which have been mapped to a reference genome and to determine if the mapped RNAs are compatible with miRNA biogenesis. miRDeep2 consists of three modules: mapper, miRDeep2, and quantifier. The mapper.pl module preprocesses the sequencing data, the miRDeep2.pl module identifies and quantifies the miRNAs, and the quantifier.pl module performs quantification and expression profiling. The sample data for both tutorials (SRR326279) represent miRNA-seq data from Illumina single-end sequencing of the cytoplasmic fraction from the human MCF-7 cell line.

Step 1: To download miRDeep2

```
-----------------------------------------------------------------
# download miRDeep2.0.07 (www.mdc-berlin.de/8551903/en/)
-----------------------------------------------------------------
```

Step 2: To download sra data and convert into FASTQ

```
-----------------------------------------------------------------
# download SRR326279.sra data from NCBI FTP service
$ wget ftp://ftp-trace.ncbi.nlm.nih.gov/sra/sra-
instant/reads/ByExp/sra/SRX%2FSRX087%2FSRX087921/
SRR326279/SRR326279.sra
# covert sra format into fastq format
$ fastq-dump SRR326279.sra
# when it is finished, you can check:
$ ls -l
# SRR326279.fastq will be produced.

-----------------------------------------------------------------
```

Step 3: To download and prepare reference files

```
-----------------------------------------------------------------
# download human hg19 genome from Illumina iGenomes
(http://support.illumina.com/sequencing/sequencing_
software/igenome.html)
$ wget ftp://igenome:G3nom3s4u@ussd-ftp.illumina.com/
Homo_sapiens/UCSC/hg19/Homo_sapiens_UCSC_hg19.tar.gz
# download miRNA precursor and mature sequences from
miRBase (http://www.mirbase.org/)
```

```
$ wget ftp://mirbase.org/pub/mirbase/CURRENT/
hairpin.fa.gz
$ wget ftp://mirbase.org/pub/mirbase/CURRENT/mature.
fa.gz
# gunzip .gz files
$ gunzip *.gz
# link human genome and bowtie index into current
working directory
$ In -s /homo.sapiens/UCSC/hg19/Sequence/
WholeGenomeFasta/genome.fa
$ In -s /homo.sapiens/UCSC/hg19/Sequence/
BowtieIndex/genome.1.ebwt
$ In -s /homo.sapiens/UCSC/hg19/Sequence/
BowtieIndex/genome.2.ebwt
$ In -s /homo.sapiens/UCSC/hg19/Sequence/
BowtieIndex/genome.3.ebwt
$ In -s /homo.sapiens/UCSC/hg19/Sequence/
BowtieIndex/genome.4.ebwt
$ In -s /homo.sapiens/UCSC/hg19/Sequence/
BowtieIndex/genome.rev.1.ebwt
$ In -s /homo.sapiens/UCSC/hg19/Sequence/
BowtieIndex/genome.rev.2.ebwt
# use mirDeep2 rna2dna.pl to substitute 'u' and 'U'
to 'T' from miRNA precursor and mature sequences
$ rna2dna.pl hairpin.fa > hairpin2.fa
$ rna2dna.pl mature.fa > mature2.fa
# when it is finished, you can check:
$ ls -l
# the following files will be produced: genome.fa,
genome.1.ebwt, genome.2.ebwt, genome.3.ebwt,
genome.4.ebwt, genome.rev.1.ebwt, genome.rev.2.ebwt,
hairpin.fa, mature.fa, hairpin2.fa and mature2.fa
```

Step 4: To extract human precursor and mature miRNA

```
# copy perl script below into hsa_edit.pl and put it
into current directory
**********************************************************
#!/usr/bin/perl
use strict;
open IN, "< hairpin2.fa";
```

```perl
open OUT,"> hairpin_hsa_dna.fa";
my $hairpin = 0;
while(my $line = <IN>){
    s/\n|\s+$//;
    if($hairpin==1){
        print OUT "$line";
        $hairpin = 0;
    }
    if($line =~/(>hsa\S+)/){
     print OUT "$line";
     $hairpin = 1;
    }
}
close IN;
close OUT;
open IN2,"< mature2.fa";
open OUT2,"> mature_hsa_dna.fa";
my $mature = 0;
while(my $line = <IN2>){
    s/\n|\s+$//;
    if($mature==1){
        print OUT2 "$line";
        $mature = 0;
    }
    if($line =~/(>hsa\S+)/){
        print OUT2 "$line";
        $mature = 1;
    }
}
close IN2;
close OUT2;
*****************************************************
# run the scripts to obtain human precursor and
mature miRNA sequences.
$ perl hsa_edit.pl
# when it is finished, you can check:
$ ls -l
# hairpin_hsa_dna.fa and mature_hsa_dna.fa will be
produced.
```

Step 5: To map reads into human genome

---

miRDeep2 mapper.pl processes the reads and maps them to the reference genome. The input file is the fastq file (SRR326279.fastq). The parameter -v outputs progress report; -q maps with one mismatch in the seed; -n overwrites existing files; -o is number of threads to use for bowtie; -u do not remove directory with temporary files; -e means input file is fastq format; -h parses to fasta format; -m collapses reads; -k clips 3′ adapter sequence AATCTCGTATGCCGTCTTCTGCTTGC; -p maps to genome; -s prints processed reads to this file (reads_collapsed.fa); -t prints read mappings to this file (reads_collapsed_vs_genome.arf).

```
$ mapper.pl SRR326279.fastq -v -q -n -o 4 -u -e -h -m -k
AATCTCGTATGCCGTCTTCTGCTTGC  -p genome -s reads_
collapsed.fa -t reads_collapsed_vs_genome.arf
# when it is finished, you can check:
$ ls -l
# reads_collapsed.fa and reads_collapsed_vs_genome.
arf will be produced.
```

Step 6: To identify known and novel miRNAs

---

miRDeep2.pl performs known and novel micoRNA identification. The input files are processed read sequences (reads_collapsed.fa), whole-genome sequences (genome.fa), mapping information (reads_collapsed_vs_genome.arf), miRNA sequences (mature_hsa_dna.fa), none (no mature sequences from other species), and miRNA precursor sequences (hairpin_hsa_dna.fa). The parameter -t is species (e.g., Human or hsa). 2>report.log pipe all progress output to report.log. The quantifier.pl module is embedded in the miRDeep2.pl module and will measure the reads count for each miRNA. The total counts will be presented in both the result.html and expression.html output files. The output pdf directory shows structure, score breakdowns, and reads signatures of known and novel miRNAs; the html webpage file (result.html) shows annotation and expression of known and novel miRNA.

```
$ miRDeep2.pl reads_collapsed.fa genome.fa reads_
collapsed_vs_genome.arf mature_hsa_dna.fa none
hairpin_hsa_dna.fa -t hsa 2>report&
# when it is finished, you can check:
$ ls -l
# result.html, expression.html and pdf directory
will be produced.
```

7.4.2 Tutorial 2: Small RNA Workbench Pipeline

Genboree (http://www.genboree.org/site/) offers a web-based platform for high-throughput sequencing data analysis using the latest bioinformatics tools. The exceRpt small RNA-seq pipeline in Genboree workbench will be used for miRNA-seq analysis based on GUI. The pipeline contains preprocessing filtering QC, endogenous alignment, and exogenous alignment. Before you start, you need to register and establish an account. We will use the same miRNA-seq sample data used in Tutorial 1. The entry page for this GUI consists of menu headings for System Network, Data, Genome, Transcriptome, Cistrome, Epigenome, Metagenome, Visualization, and Help. Each of these headings will have drop down menus. There are also four main boxes for experimental set up and analysis, including Data Selector, Details, Input Data, and Output Targets.

> **Step 1:** *Create new group in Genboree.* At first, drag Data Selector **genboree.org** into **Output Targets** box, click **System/Network -> Groups -> Create Group**, type in **miRNA-seq example** as Group Name and **Genboree miRNA-seq example** as Description. Click **Submit**. Job Submission Status will assign a job id. Click **OK**. Click Data Selector **Refresh** and click Output Targets **Remove** button to inactive **genboree.org**. Step 1 is necessary to establish a working group to analyze miRNA

> **Step 2:** *Create new database in Genboree.* Drag Data Selector **miRNA-seq example** into **Output Targets**, click **Data -> Databases -> Create**

Database, set **Template: Human (hg19)** as Reference Sequence, type in **miRNA-seq** as Database Name, and type in **miRNA-seq data analysis** as Description. **Homo sapiens** as Species and **hg19** as Version should be automatically filled. Click **Submit**. Job Submission Status will assign a job id. Click **OK**. Click Data Selector **Refresh** and click Output Targets **Remove** button to inactive the **miRNA-seq example** target.

Step 3: *Transfer SRR326279.fastq data into Genboree FTP server.* Click Data Selector **miRNA-seq example -> Databases**, drag **miRNA-seq** into **Output Targets**. And click **Data -> Files -> Transfer File**, click **Choose File** button to select **SRR326279.fastq** and click **Open**. Refer to **Step 2** in the **miRDeep2 tutorial** on how to download the **SRR326279.fastq** file to your computer. Set **Test** as Create in SubFolder and **SRR326279 for miRNA-seq example** as File Description and click **Submit.** Job Submission Status will assign a job id. Click **OK**. Click Data Selector **Refresh** and click Output Targets **Remove** to inactive the **miRNA-seq** target.

Step 4: *Run exceRpt small RNA-seq pipeline.* Now that the experimental group has been established and the reference genome and sequencing data have been uploaded, the analysis step can be initiated. Click Data Selector **miRNA-seq example -> Databases -> miRNA-seq -> Files -> Test**, and drag **SRR326279.fastq** into **Input Data** and database **miRNA-seq** into **Output Targets**. Multiple *.fastq sample files can be submitted together. To analyze additional *.fastq files for the same experiment, proceed with **Step 3**; it is not necessary to repeat **Steps 1** and **2**. Then click **Transcriptome -> Analyze Small RNA-Seq Data -> exceRpt small RNA-seq Pipeline**. Set the parameters for miRNA-seq analysis in **Tool Settings**. Enter **AATCTCGTATGCCGTCTTCTGCTTGC** as 3′ Adapter Sequence and choose **Endogenous-only** as small RNA Libraries. **Defaults** are used for other parameters. Click **Submit**. Job Submission Status will provide a job id for this analysis. Click **OK**. This step will take several hours to complete and is dependent upon the number of samples submitted for analysis. Once the files have been submitted for analysis, the program can be closed.

Step 5: *Download analysis results.* An e-mail notice will be sent when the analysis is completed. Log-in to your account and click Data Selector

miRNA-seq example -> Databases -> miRNA-seq -> Files -> small-RNAseqPipeline -> smallRNA-seq Pipeline -> processed Results. A panel of 15 different results will be reported (e.g., mapping summary, miRNA count, piRNA count, and tRNA count). If you want to download those files, click the **file** followed by Details **Click to Download File**.

BIBLIOGRAPHY

Desvignes, T., Beam, M. J., Batzel, P., Sydes, J., and Postlethwait, J. H. (2014). Expanding the annotation of zebrafish microRNAs based on small RNA sequencing. *Gene, 546*(2), 386–389. doi:10.1016/j.gene.2014.05.036.

Eminaga, S., Christodoulou, D. C., Vigneault, F., Church, G. M., and Seidman, J. G. (2013). Quantification of microRNA expression with next-generation sequencing. *Curr Protocol Mol Biol,* Chapter 4, Unit 4 17. doi:10.1002/0471142727. mb0417s103.

Fatima, A., Waters, S., O'Boyle, P., Seoighe, C., and Morris, D. G. (2014). Alterations in hepatic miRNA expression during negative energy balance in postpartum dairy cattle. *BMC Genomics, 15*, 28. doi:10.1186/1471-2164-15-28.

Friedlander, M. R., Chen, W., Adamidi, C., Maaskola, J., Einspanier, R., Knespel, S., and Rajewsky, N. (2008). Discovering microRNAs from deep sequencing data using miRDeep. *Nat Biotechnol, 26*(4), 407–415. doi:10.1038/nbt1394.

Friedlander, M. R., Mackowiak, S. D., Li, Na., Chen, W., and Rajewsky, N. (2011). miRDeep2 accurately identifies known and hundreds of novel microRNA genes in seven animal clades. *Nucl Acids Res, 40*(1), 37–52. doi:10.1093/nar/gkr688.

Friedlander, M. R., Lizano, E., Houben, A. J., Bezdan, D., Banez-Coronel, M., Kudla, G., Mateu-Huertas, E et al. (2014). Evidence for the biogenesis of more than 1,000 novel human microRNAs. *Genome Biol, 15*(4), R57. doi:10.1186/gb-2014-15-4-r57.

Gomes, C. P., Cho, J. H., Hood, L., Franco, O. L., Pereira, R. W., and Wang, K. (2013). A review of computational tools in microRNA discovery. *Front Genet, 4*, 81. doi:10.3389/fgene.2013.00081.

Gunaratne, P. H., Coarfa, C., Soibam, B., and Tandon, A. (2012). miRNA data analysis: Next-gen sequencing. *Methods Mol Biol, 822*, 273–288. doi:10.1007/978-1-61779-427-8_19.

Ha, M., and Kim, V. N. (2014). Regulation of microRNA biogenesis. *Nat Rev Mol Cell Biol, 15*(8), 509–524. doi:10.1038/nrm3838.

Hackenberg, M., Sturm, M., Langenberger, D., Falcon-Perez, J. M., and Aransay, A. M. (2009). miRanalyzer: A microRNA detection and analysis tool for next-generation sequencing experiments. *Nucleic Acids Res, 37*(Web Server issue), W68–W76. doi:10.1093/nar/gkp347.

Han, Y., Chen, J., Zhao, X., Liang, C., Wang, Y., Sun, L.,. . . Cai, Z. (2011). MicroRNA expression signatures of bladder cancer revealed by deep sequencing. *PLoS One, 6*(3), e18286. doi:10.1371/journal.pone.0018286.

Higuchi, C., Nakatsuka, A., Eguchi, J., Teshigawara, S., Kanzaki, M., Katayama, A., Yamaguchi, S et al. (2015). Identification of circulating miR-101, miR-375 and miR-802 as biomarkers for type 2 diabetes. *Metabolism, 64*(4), 489–497. doi:10.1016/j.metabol.2014.12.003.

Kuchen, S., Resch, W., Yamane, A., Kuo, N., Li, Z., Chakraborty, T., Wei, L et al. (2010). Regulation of microRNA expression and abundance during lymphopoiesis. *Immunity, 32*(6), 828–839. doi:10.1016/j.immuni.2010.05.009.

Liu, N., and Olson, E. N. (2010). MicroRNA regulatory networks in cardiovascular development. *Dev Cell, 18*(4), 510–525. doi:10.1016/j.devcel.2010.03.010.

Londin, E., Loher, P., Telonis, A. G., Quann, K., Clark, P., Jing, Y., Hatzimichael E. et al. (2015). Analysis of 13 cell types reveals evidence for the expression of numerous novel primate- and tissue-specific microRNAs. *Proc Natl Acad Sci U S A, 112*(10), E1106–E1115. doi:10.1073/pnas.1420955112.

Londin, R., Gan, I., Modai, S., Sukacheov, A., Dror, G., Halperin, E., and Shomron, N. (2010). miRNAkey: A software for microRNA deep sequencing analysis. *Bioinformatics, 26*(20), 2615–2616. doi:10.1093/bioinformatics/btq493.

Ma, J., Mannoor, K., Gao, L., Tan, A., Guarnera, M. A., Zhan, M., Shetty, A et al. (2014). Characterization of microRNA transcriptome in lung cancer by next-generation deep sequencing. *Mol Oncol, 8*(7), 1208–1219. doi:10.1016/j.molonc.2014.03.019.

Marti, E., Pantano, L., Banez-Coronel, M., Llorens, F., Minones-Moyano, E., Porta, S., Sumoy, L et al. (2010). A myriad of miRNA variants in control and Huntington's disease brain regions detected by massively parallel sequencing. *Nucleic Acids Res, 38*(20), 7219–7235. doi:10.1093/nar/gkq575.

Ronen, R., Gan, I., Modai, S., Sukacheov, A., Dror, G., Halperin, E., and Shomron, N. (2010). miRNAkey: a software for microRNA deep sequencing analysis. *Bioinformatics, 26*(20), 2615–2616. doi:10.1093/bioinformatics/btq493.

Shimizu, C., Kim, J., Stepanowsky, P., Trinh, C., Lau, H. D., Akers, J. C., Chen, C et al. (2013). Differential expression of miR-145 in children with Kawasaki disease. *PLoS One, 8*(3), e58159. doi:10.1371/journal.pone.0058159.

Somel, M., Liu, X., Tang, L., Yan, Z., Hu, H., Guo, S., Jian, X et al. (2011). MicroRNA-driven developmental remodeling in the brain distinguishes humans from other primates. *PLoS Biol, 9*(12), e1001214. doi:10.1371/journal.pbio.1001214.

Sun, Z., Evans, J., Bhagwate, A., Middha, S., Bockol, M., Yan, H., and Kocher, J. P. (2014). CAP-miRSeq: A comprehensive analysis pipeline for microRNA sequencing data. *BMC Genom, 15*, 423. doi:10.1186/1471-2164-15-423.

Wang, J., Yu, J. T., Tan, L., Tian, Y., Ma, J., Tan, C. C., Wang, H. F et al. (2015). Genome-wide circulating microRNA expression profiling indicates biomarkers for epilepsy. *Sci Rep, 5*, 9522. doi:10.1038/srep09522.

Wienholds, E., and Plasterk, R. H. (2005). MicroRNA function in animal development. *FEBS Lett, 579*(26), 5911–5922. doi:10.1016/j.febslet.2005.07.070.

Xia, R., Zhu, H., An, Y. Q., Beers, E. P., and Liu, Z. (2012). Apple miRNAs and tasiRNAs with novel regulatory networks. *Genome Biol, 13*(6), R47. doi: 10.1186/gb-2012-13-6-r47.

## CHAPTER 8

Methylome-Seq Data Analysis

Chengpeng Bi

CONTENTS

8.1 INTRODUCTION

Methylation of cytosines across genomes is one of the major epigenetic modifications in eukaryotic cells. DNA methylation is a defining feature of mammalian cellular identity and is essential for normal development. Single-base resolution DNA methylation is now routinely being decoded by combining high-throughput sequencing with sodium bisulfite conversion, the gold standard method for the detection of cytosine DNA methylation. Sodium bisulfite is used to convert unmethylated cytosine to uracil and ultimately thymine, and thus, the treatment can be used to detect the methylation state of individual cytosine nucleotides. In other words, a methylated cytosine will not be impacted by the treatment; however, an unmethylated cytosine is most likely converted to a thymine. DNA methylation occurs predominantly at cytosines within CpG (cytosine and guanine separated by only one phosphate) dinucleotides in the mammalian genome, and there are over 28 million CpG sites in the human genome. High-throughput sequencing of bisulfite-treated DNA molecules allows resolution of the methylation state of every cytosine in the target sequence, at single-molecule resolution, and is considered the *gold standard* for DNA methylation analysis. This bisulfite-sequencing (BS-Seq) technology allows scientist to investigate the methylation status of each of these CpG sites genome-wide. A methylome for an individual cell type is such a gross mapping of each DNA methylation status across a genome.

Coupling bisulfite modification with next-generation sequencing (BS-Seq) provides epigenetic information about cytosine methylation at single-base resolution across the genome and requires the development of bioinformatics pipeline to handle such a massive data analysis. Because of the cytosine conversions, we need to develop bioinformatics tools specifically suited for the volume of BS-Seq data generated. First of all, given the methylation sequencing data, it is necessary to map the derived sequences back to the reference genome and then determine their methylation status on each cytosine residue. To date, several BS-Seq alignment tools have been developed. BS-Seq alignment algorithms are used to estimate percentage methylation at specific CpG sites (methylation calls), but also provide the ability to call single nucleotide and small indel variants as well as copy number and structural variants. In this chapter, we will focus on the challenge presented by methylated sequencing alignment and methylation status. There are basically two strategies used to perform methylation sequencing alignment: (1) wild-card matching approaches, such as BSMAP, and (2) three-letter aligning algorithms, such as Bismark. Three-letter alignment is one of the most popular approaches described in the literature. It involves converting all cytosine to thymine residues on a forward stand, and guanine to adenine residues on its reverse stand. Such a conversion is applied to both reference genome and short reads, and then followed by mapping the converted reads to the converted genome using a short-read aligner such as Bowtie. Either gapped or ungapped alignment can be used, depending on the underlying short-read alignment tool.

Bismark is one of the most frequently used methylation mapping pipelines that implement the three-letter approach. It consists of a set of tools for the time-efficient analysis of BS-Seq data. Bismark simultaneously aligns bisulfite-treated reads to a reference genome and calls cytosine methylation status. Written in Perl and run from the command line, Bismark maps bisulfite-treated reads using a short-read aligner, either Bowtie1 or Bowtie2. For presentation purposes, we will use Bismark together with Bowtie2 to demonstrate the process for analysis of methylation data.

8.2 APPLICATION

DNA methylation is an epigenetic mark fundamental to developmental processes including genomic imprinting, silencing of transposable elements and differentiation. As studies of DNA methylation increase in scope, it has become evident that methylation is deeply involved in regulating gene expression and differentiation of tissue types and plays critical roles in pathological processes resulting in various human diseases. DNA methylation patterns can be inherited and influenced by the environment, diet, and aging, and disregulated in diseases. Although changes in the extent and pattern of DNA methylation have been the focus of numerous studies investigating normal development and the pathogenesis disease, more recent applications involve incorporation of DNA methylation data with other *-omic* data to better characterize the complexity of interactions at a *systems* level.

8.3 DATA ANALYSIS OUTLINE

The goal of DNA methylation data analysis is to determine if a site containing C is methylated or not across a genome. One has to perform high-throughput sequencing (BS-Seq) of converted short reads and then align each such read back onto the reference human genome. This kind of alignment is a special case of regular short-read alignment.

For example, given a set of short reads in a FASTQ file from a next-generation sequencing platform such as Illumina sequencing machine, the first step is to perform quality control of reads by running the FastQC program, for detail, refer to the FastQC website (http://www.bioinformatics.babraham.ac.uk/projects/fastqc/). The QC-passed reads are then subject to mapping onto a reference genome. For the human genome, the reference sequence can be downloaded from the University of California, Santa Cruz (UCSC; http://genome.ucsc.edu), or Ensembl. To prepare for mapping of short reads, the reference genomic sequence is subject to base conversion computationally, and two separate converted genomes must be considered: one in which C to T conversion on forward strand, and another with G to A conversion on the reverse strand. Similarly, bisulfite-treated short reads are also subject to two kind of conversions, and each converted read is mapped to its associated reference sequence whereby one can determine if a position is methylated or not. After read mapping, a potential methylated site from all the aligned short reads can be summarized, each having the same genomic location, that is, summarizing them on one row: counting how many methylated and how many unmethylated from all reads at the same site. Figure 8.1 exhibits the flowchart of how the procedures are performed.

For the methylation pipeline presented in Figure 8.1, Bismark is applied and is used together with Bowtie in this flowchart. The working procedure of Bismark begins with read conversion, in which the sequence reads are first transformed into completely bisulfite-converted forward (C->T) and its cognate reverse read (G->A conversion of the reverse strand) versions, before they are aligned to similarly converted versions of the genome (also C->T and G->A converted). Bismark aligns all four possible alignments for each read and pick the best alignment, that is, sequence reads that produce a unique best alignment from the four alignment processes against the bisulfite genomes (which are running in parallel) are then compared to the normal genomic sequence, and the methylation state of all cytosine positions in the read is inferred. For use with Bowtie1, a read is considered to align uniquely if a single alignment exists that has with fewer mismatches to the genome than any other alternative alignment if any. For Bowtie2, a read is considered to align uniquely if an alignment has a unique best alignment score. If a read produces several alignments with the same number of mismatches or with the same alignment score,

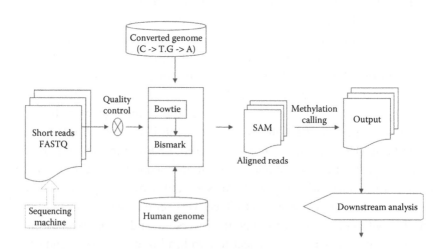

FIGURE 8.1 Flowchart of DNA methylation data analysis.

a read (or a read-pair) is discarded altogether. Finally, Bismark output its calling results in SAM format with several new extended fields added and also throw away a few fields from original Bowtie output.

After methylation calling on every sites detected, we need to determine methylation status based on a population of the same type of cells or short reads on each cytosine sites. There will be two alternative statuses to appear on each site: either methylated or unmethylated due to random errors for various reasons, see a demonstration in Figure 8.2a. Therefore, statistical method is needed to determine if a site is really methylated or not. Figure 8.2b demonstrates this scenario. Although bisulfite treatment is used to check if a base C is methylated or not, there are a lot of reasons that may give different outcomes, and we want to statistically test which outcome is the dominant one and conclude a true methylation status on each site. In Figure 8.2a, there are two CpG sites in the DNA sequence, the first C is methylated and not converted after bisulfite treatment as in highlighted area, the second C is not methylated and it is converted to T. Therefore, after bisulfite treatment, all sites with methylated cytosine are most likely not impacted, whereas unmethylated Cs are most probably

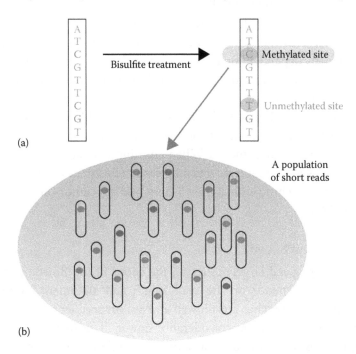

FIGURE 8.2 Population of short reads in DNA methylation. (a) Bisulphite treatment of a short read. (b) A population of treated short reads.

converted to Ts. In Figure 8.2b, there is a population of such cells or reads with experimental bias, that is, on the same site there may be two methylation results due to various reasons. This is a typical Bernoulli experiment with two possible outcomes: methylated or not. In this demonstration, there are 5 reads showing unmethylated at a site, whereas 15 reads display methylated on the same site, so the frequency of methylation on the site is 3/4, and unmethylated is 1/4. Therefore, the site detected is significantly methylated ($p < .05$).

8.4 STEP-BY-STEP TUTORIAL ON BS-SEQ DATA ANALYSIS

8.4.1 System Requirements

A minimum knowledge of Linux/Unix system is required to a pipeline user. The Linux/Unix system has already equipped with Perl language with which Bismark is written, and GNU GCC compiler is needed to compile the source code of Bowtie2, which is written in C/C++ language. Both Perl and GCC are free software and publicly available.

8.4.2 Hardware Requirements

As reported, Bismark holds the reference genome in memory while running Bowtie, with four parallel instances of the program. The memory usage is largely dependent on the size of the reference genome and BS-Seq data. For a large eukaryotic genome such as human genome, a typical memory usage of around 16 GB is needed. It is thus recommended running Bismark on a Linux/Unix machine with 5 CPU cores and 16 GB RAM. The memory requirements of Bowtie2 are a little larger than Bowtie1 if allowing gapped alignments. When running Bismark combined with Bowtie2, the system requirements may need to be increased, for example, a Linux/Unix machine with at least 5 cores and its memory size of at least 16 GB of RAM.

8.4.3 Alignment Speed

The alignment speed largely relies on the speed of the Bowtie program, which in turn depends on the read length and alignment parameters used. If many mismatches are allowed and a short seed length is used, the alignment process will be considerably slower. If near-perfect matches are required, Bowtie1 can align around 5–25 million sequences per hour. Bowtie2 is often much faster than Bowtie1 under similar run conditions.

8.4.4 Sequence Input

Bismark is a pipeline specified for the alignment of bisulfite-treated reads. The reads may come either from whole-genome shotgun BS-Seq (WGSBS) or from reduced-representation BS-Seq (RRBS). The input read sequence file can be in the format of either FastQ or FastA. The sequences can be single-end or paired-end reads. The input files can be in the format of either uncompressed plain text or gzip-compressed text (using the .gz file extension). The short-read length in each sequences can be different. The reads can be coming from either directional or non-directional BS-Seq libraries.

8.4.5 Help Information

A full list of alignment modes can be found at http://www.bioinformatics. babraham.ac.uk/projects/bismark/Bismark_alignment_modes.pdf.

In addition, Bismark retains much of the flexibility of Bowtie1/ Bowtie2.

8.4.6 Tutorial on Using Bismark Pipeline

A detailed tutorial on how to download and install the software used and prepare reference genome sequence is provided in the following sections. Examples describing the aligning and mapping procedures are also provided.

Step 1: *Download of Bismark methylation pipeline as well as Bowtie short-read aligner.* To get the current version of Bismark v0.14.0, you may go to the downloading website: http://www.bioinformatics.babraham. ac.uk/projects/download.html#bismark. The compressed filename downloaded is bismark_v0.14.0.tar.gz. The zipped file should be installed on a Linux/Unix machine, for example, in my home directory: /home/cbi/, and then unpack the zipped file by executing the following Linux/Unix command in the current directory such as/ home/cbi/:

```
[cbi@head ~]$ tar zxvf bismark_v0.14.0.tar.gz
```

For a full list of options while using Bismark, run the following:

```
[cbi@head bismark_v0.14.0]$ ./bismark --help
```

Bismark will be automatically installed onto /home/cbi/bismark_ v0.14.0, and you simply go there by typing the command: cd bis-mark_v0.14.0. There are two important programs found: one is bismark_genome_preparation, and another is bismark. We will use these two programs soon.

Because bismark is a pipeline, which means it relies on another core short-read aligning program called bowtie to perform methylated sequence alignment, we have to download and install Bowtie software before running bismark. We are going to download the fast and accurate version of Bowtie2 version 2.2.5 from the public website: http://sourceforge.net/projects/bowtie-bio/files/bowtie2/2.2.5/. The zipped filename is bowtie2-2.2.5-source.zip, and then, we need to unzip the file as follows:

```
[cbi@head ~]$ unzip bowtie2-2.2.5-source.zip
```

Then, we go to the bowtie2 directory by typing: cd bowtie2-2.2.5 and then type the command 'make' to compile and install the software. Note that GCC compiler should be available in your Linux/ Unix machine or server, if not, you need to ask your system administrator to install it.

Step 2: *Download of human genome sequence.* We may go to the ENSEMBL site to download the human genome: ftp://ftp.ensembl.org/pub/ release-78/fasta/homo_sapiens/dna/. Other sites could be from NCBI or UCSC genome browser. After that, you need to transfer the genome sequence into the target Linux/Unix machine, better putting it in a common use site to be shared with other users. For example, we put the human genome to the reference folder as /data/scratch2/ hg38/. We create the genome folder under the directory /data/ scratch2 as follows:

```
[cbi@head ~]$mkdir /data/scratch2/hg38
```

Step 3: *Preparation of reference genome and Bowtie indexing libraries.* The goal of this step is to prepare reference indexing libraries in order to perform read alignment for bowtie. This Perl script (bismark_ genome_preparation) needs to be run only once to prepare the genome of your interest for bisulfite-treated short-read alignments.

First, we need to create a directory containing the genome downloaded as mentioned above. Note that the Perl script `bismark_genome_preparation` currently expects FASTA files in this folder (with either .fa or .fasta extension, single combined or multiple chromosome sequence files per genome). Bismark will automatically create two individual subfolders under the genome directory, one for a C->T converted reference genome and the other one for the G->A converted reference genome. After creating C->T and G->A versions of the genome, they will be indexed in parallel using the bowtie indexer `bowtie-build` (or `bowtie2-build`). It will take quite a while for Bowtie to finish preparing both C->T and G->A genome indices. This preparation is done once for all. Please note that Bowtie1 and Bowtie2 indexes are very different and not compatible; therefore, you have to create them separately. To create a genome index for use with Bowtie2, the option `--bowtie2` needs to be included in the command line as well.

For the BS-Seq short-read alignment, we need to prepare indices for the reference genome by running the following command in bowtie2 mode:

```
[cbi@head~]$ /home/cbi/bismark0.14.0/bismark_
genome_preparation --bowtie2 --path_to_bowtie /home/
cbi/bowtie2-2.2.5 --verbose /data/scratch2/hg38/
```

The above step will create two indexing libraries in order to align the methylated short reads by bowtie2. The indexing data sets will be put under the reference genome folder auto-created as `Bisulfite_Genome` under which there are two subfolders to store the Bowtie2 indexing libraries: `CT_conversion` and `GA_conversion`.

Step 4: *Running Bismark.* This step is the actual bisulfite-treated short-read alignment by the bowtie program and methylation calling by Bismark as the post-alignment process. Bismark asks the user to provide two key parameters: (1) The directory contains the reference genome as shown above. Note that we assume there are Bowtie1 or Bowtie2 indices already built under the genome directory, which means that this folder must contain the original genome (a fasta-formatted single or multiple files per genome), and the two bisulfite genome subdirectories already generated as above. This is required, otherwise the alignment will not work. (2) A single or multiple sequence files consist of all bisulfite-treated short reads in either FASTQ or FASTA format. All other information is optional.

In the current version, it is required that the current working directory contains the short-read sequence files to be aligned. For each short-read sequence file or each set of paired-end sequence files, Bismark produces one alignment as well as its methylation calling information as output file. Together, a separate report file describing alignment and methylation calling statistics also provides for user's information on alignment efficiency and methylation percentages.

Bismark can run with either Bowtie1 or Bowtie2. It is defaulted to Bowtie1. If Bowtie2 is needed, one has to specify as `--bowtie2`. Bowtie1 is run default as `--best` mode. Bowtie1 uses standard alignments allowing up to 2 mismatches in the seed region, which is defined as the first 28 bp by default. These parameters can be modified using the options `-n` and `-l`, respectively. We recommend the default values for a beginner.

When Bismark calls Bowtie2, it uses its standard alignment settings. This means the following: (1) It allows a multi-seed length of 20 bp with 0 mismatches. These parameters can be modified using the options `-L` and `-N`, respectively. (2) It reports the best of up to 10 valid alignments. This can be set using the `–M` parameter. (3) It uses the default minimum alignment score function `L,0,-0.2`, i.e., $f(x) = 0 + -0.2 * x$, where x is the read length. For a read of 75 bp, this would mean that a read can have a lowest alignment score of −15 before an alignment would become invalid. This is roughly equal to 2 mismatches or ~2 indels of 1–2 bp in the read.

Bisulfite treatment of DNA and subsequent polymerase chain reaction (PCR) amplification can give rise to four (bisulfite converted) strands for a given locus. Depending on the adapters used, BS-Seq libraries can be constructed in two different ways: (1) If a library is directional, only reads which are (bisulfite converted) versions of the original top strand or the original bottom strand will be sequenced. By default, Bismark performs only 2 read alignments to the original strands, called `directional`. (2) Alternatively, BS-Seq libraries can be constructed so that all four different strands generated in the BS-PCR can and will end up in the sequencing library with roughly the same likelihood.

In this case, all four strands can produce valid alignments, and the library is called non-directional. While choosing --non_ directional, we ask Bismark to use all four alignment outputs, and it will double the running time as compared to directional library.

A methylation data file is often in FASTQ format; for example, we download a testing file from NCBI website as follows:

```
[cbi@head~]$ wget
```

```
"ftp://hgdownload.cse.ucsc.edu/goldenPath/hg18/
encodeDCC/wgEncodeYaleChIPseq/
wgEncodeYaleChIPseqRawDataRep1Gm12878NfkbTnfa.
fastq.gz"
```

Then, we unzip the fastq file and rename it as test.fastq for simplicity as follows:

```
[cbi@head~]$ gunzip
wgEncodeYaleChIPseqRawDataRep1Gm12878NfkbTnfa.
fastq.gz
[cbi@head~]$mv
wgEncodeYaleChIPseqRawDataRep1Gm12878NfkbTnfa.
fastq test.fastq
```

Now the sequence file test.fastq is in current working folder, and we run Bismark to align all the short reads in the file unto converted reference genomes as prepared in step 3. The following command is executed:

```
[cbi@head~]$ /home/cbi/bismark0.14.0/bismark
--bowtie2 --non_directional --path_to_bowtie /home/
cbi/bowtie2-2.2.5 /data/scratch2/hg38/ test.fastq
```

The above command will produce two output files: (a) test.fastq_ bismark_bt2.bam, holding information on all short reads aligned, plus methylation calling strings and reference and read conversions used; (b) test.fastq_bismark_bt2_SE_report.txt, holding information on alignment and methylation summary. We use the following command to generate full plain text in SAM format from the binary formatted BAM file:

```
[cbi@head~]$samtools view  -h test.fastq_bismark_
bt2.bam    >test.fastq_bismark_bt2.sam
```

Note that you have to ask your system administrator to install samtools before you run the above command. If it is pair-ended sequencing, for example, a pair of read files given as test1.fastq and test2.fastq, we execute the following:

```
[cbi@head~]$ /home/cbi/bismark0.14.0/bismark
--bowtie2 --non_directional --path_to_bowtie
/home/cbi/bowtie2-2.2.5 /data/scratch2/hg38/ -1
test1.fastq -2 test2.fastq
```

By default, the most updated version of Bismark will generate BAM output for all alignment modes. Bismark can generate a comprehensive alignment and methylation calling output file for each input file or set of paired-end input files. The sequence base-calling qualities of the input FastQ files are also copied into the Bismark output file as well to allow filtering on quality thresholds if needed. Note that the quality values are encoded in Sanger format (Phred 33 scale). If the input format was in Phred64 or the old Solexa format, it will be converted to Phred 33 scale.

The single-end output contains the following important information in SAM format: (1) seq-ID, (2) alignment strand, (3) chromosome, (4) start position, (5) mapping quality, (6) extended CIGAR string, (7) mate reference sequence, (8) 1-based mate position, (9) inferred template length, (10) original bisulfite read sequence, (11) equivalent genomic sequence (+2 extra bp), (12) query quality, (13) methylation call string (XM:Z), (14) read conversion (XR:Z), and (15) genome conversion (XG:Z). Here is an example from the output file test.fastq_bismark_bt2.sam:

```
FC30WN3HM_20090212:3:1:212:1932 16      16
59533920        42       28M      *       0       0
GTATTTGTTTTCCACTAGTTCAGCTTTC    [[Z[]Z]Z[]][]]
[[]][]]]]]]][]  NM:i:0  MD:Z:28 XM:Z:H.....H...
.......H....X..... XR:Z:CT XG:Z:GA
```

If a methylation call string contains a dot ".", it means not involving a cytosine. Otherwise, it contains one of the following letters for the three different cytosine methylation contexts:

```
z unmethylated C in CpG context (lower case means
unmethylated)
Z methylated C in CpG context (upper case means
methylated)
x unmethylated C in CHG context
X methylated C in CHG context
h unmethylated C in CHH context
H methylated C in CHH context
u unmethylated C in Unknown context (CN or CHN)
U methylated C in Unknown context (CN or CHN)
```

In fact, the methylation output in SAM format generated from Bismark provides opportunity for those users who can write Perl or other scripts to code their own scripts to extract and aggregate methylation status across genome for each individual samples. If this is the case, you can skip step 5.

Step 5: *Methylation calling.* The goal of this step is to aggregate methylation status for each site across genome. Clearly, most investigators are often interested in methylation sites on CpG context. Besides the two programs used as above, Bismark also provides users with a supplementary Perl script called `bismark_methylation_extractor`, which operates on Bismark output results and extracts the methylation calling information for every single C, methylated or not. After processing by the extractor, the position of every single C will be written out to a new output file, together with one of three contexts: CpG, CHG, or CHH. The methylated Cs will be labeled as forward reads (+) and non-methylated Cs as reverse reads (−). The resulting files can be imported into a genome viewer such as SeqMonk (using the generic text import filter) for further analysis. The output of the methylation extractor can be also transformed into a bedGraph file using the option `--bedGraph`. This step can also be accomplished from the methylation extractor output using the stand-alone script `bismark2bedGraph`. As its default option, the `bismark_methylation_extractor` will produce a strand-specific output which will use the following abbreviations in the output file name to indicate the strand the alignment came from one of four possible situations: OT, original top strand; CTOT, complementary to original top strand; OB, original bottom strand; and CTOB, complementary to original bottom strand.

A typical command to extract context-dependent (CpG/CHG/CHH) methylation could look like this:

```
[cbi@head~]$/home/cbi/bismark0.14.0/bismark_
methylation_extractor -s --comprehensive test.
fastq_bismark_bt2.sam
```

This will produce three output files each having four source strands (STR takes either OT, OB, CTOT, or CTOB) given as follows:

(a) CpG_STR_context_test.fastq_bismark_bt2.txt
(b) CHG_STR_context_test.fastq_bismark_bt2.txt
(c) CHH_STR_context_test.fastq_bismark_bt2.txt

The methylation extractor output has the following items (tab separated): (1) seq-ID, (2) methylation state (+/−), (3) chromosome number, (4) start position (= end position), and (5) methylation calling. Examples for cytosines in CpG context (Z/z) are

```
FC30WN3HM_20090212:3:1:214:1947 +          18
10931943            Z
FC30WN3HM_20090212:3:1:31:1937  +          6
77318837            Z
```

A typical command including the optional --bedGraph --counts output could look like this:

```
[cbi@head~]$/home/cbi/bismark0.14.0/bismark_
methylation_extractor -s --bedGraph --counts
--buffer_size 10G test.fastq_bismark_bt2.sam
```

The output data are in the current folder named as test.fastq_bismark_bt2.bedGraph. The content is something like this (first column is chromosome number, second is start position, third is end position, and last is methylation percentage):

```
track type=bedGraph
18      267928  267929  0
18      268002  268003  100
18      268005  268006  100
18      268008  268009  100
```

A typical command including the optional genome-wide cytosine methylation report could look like this:

```
[cbi@head~]$/home/cbi/bismark0.14.0/bismark_
methylation_extractor -s --bedGraph --counts
--buffer_size 10G --cytosine_report --genome_folder
/data/scratch2/hg38/ test.fastq_bismark_bt2.sam
```

The above output is stored in the file: `test.fastq_bismark_bt2.CpG_report.txt`, from where we extract part of data like this:

chr#	position	strand	#methyl	#unmethyl	CG	tri-nucleotide
5	49657477	-	33	2	CG	CGA
2	89829453	+	29	1	CG	CGT
10	41860296	-	81	7	CG	CGG

Step 6: *Testing if a site is methylated.* The above data with counts of methylated and unmethylated for each sites can be uploaded into a spreadsheet and perform *t*-test or other methods available and check if a site is significantly methylated.

ACKNOWLEDGMENT

The author thanks Dr. J. Steve Leeder for his comments and for proofreading the manuscript.

BIBLIOGRAPHY

1. Ziller MJ, Gu H, Muller F et al. Charting a dynamic DNA methylation landscape of the human genome. *Nature* 2013; 500:477–481.
2. Lister R et al. Human DNA methylomes at base resolution show widespread epigenomic differences. *Nature* 2009; 462:315–322.
3. Pelizzola M and Ecker JR. The DNA methylome. *FEBS Lett.* 2011; 585(13):1994–2000.
4. Langmead B, Trapnell C, Pop M, and Salzberg SL. Ultrafast and memory-efficient alignment of short DNA sequences to the human genome. *Genome Biol.* 2009; 10:R25.

5. Krueger F and Andrews SR. Bismark: A flexible aligner and methylation caller for Bisulfite-Seq applications. *Bioinformatics* 2011; 27:1571–1572.
6. Otto C, Stadler PF, and Hoffmann S. Fast and sensitive mapping of bisulfite-treated sequencing data. *Bioinformatics* 2012; 28(13):1689–1704.
7. Xi Y and Li W. BSMAP: Whole genome bisulfite sequence MAPping program. *BMC Bioinform.* 2009; 10:232.
8. Jones PA. Functions of DNA methylation: Islands, start sites, gene bodies and beyond. *Nat. Rev. Genet.* 2012; 13:484–492.
9. Li Y and Tollefsbol TO. DNA methylation detection: Bisulfite genomic sequencing analysis. *Methods Mol. Biol.* 2011; 791:11–21.
10. Stirzaker C, Taberlay PC, Statham AL, and Clark SJ. Mining cancer methylomes: Prospects and challenges. *Trends Genet.* 2014; 30(2):75–84.

## CHAPTER 9

ChIP-Seq Data Analysis

Shui Qing Ye, Li Qin Zhang, and Jiancheng Tu

CONTENTS

9.1 INTRODUCTION

Chromatin immunoprecipitation sequencing (ChIP-seq) is a method to combine chromatin immunoprecipitation with massively parallel DNA sequencing to identify the binding sites of DNA-associated proteins such as transcription factors (TFs), polymerases and transcriptional machinery, structural proteins, protein modifications, and DNA modifications. ChIP-seq can be used to map global binding sites precisely and cost effectively for any protein of interest. TFs and other chromatin-associated proteins are essential phenotype-influencing mechanisms. Determining how proteins interact with DNA to regulate gene expression is essential for fully understanding many biological processes and diseases states.

This epigenetic information is complementary to genotypes and expression analysis.

Traditional methods such as electrophoresis gel mobility shift and DNase I footprinting assays have successfully identified TF binding sites and specific DNA-associated protein modifications and their roles in regulating specific genes, but these experiments are limited in scale and resolution. This limited utility has sparked the development of chromatin immunoprecipitation with DNA microarray (ChIP-chip) to identify interactions between proteins and DNA in larger scales. However, with the advent of lower cost and higher speed next-generation DNA sequencing technologies, ChIP-seq is gradually replacing ChIP-chip as the tour de force for the detection of DNA-binding proteins on a genome-wide basis. The ChIP-seq technique usually involves fixing intact cells with formaldehyde, a reversible protein–DNA cross-linking agent that serves to fix or preserve the protein–DNA interactions occurring in the cell. The cells are then lysed and chromatin fragments are isolated from the nuclei by sonication or nuclease digestion. This is followed by the selective immunoprecipitation of protein–DNA complexes by utilizing specific protein antibodies and their conjugated beads. The cross-links are then reversed, and the immunoprecipitated and released DNA is subjected to next-generation DNA sequencing before the specific binding sites of the probed protein are identified by a computation analysis.

Over ChIP-chip, ChIP-seq has advantages of hundredfold lower DNA input requirements, no limitation on content available on arrays, more precise position resolution, and higher quality data. Of note is that the ENCyclopedia Of DNA Elements (ENCODE) and the Model Organism ENCyclopedia Of DNA Elements (modENCODE) consortia have performed more than a thousand individual ChIP-seq experiments for more than 140 different factors and histone modifications in more than 100 cell types in four different organisms (*D. melanogaster*, *C. elegans*, mouse, and human), using multiple independent data production and processing pipelines. ChIP-seq is gaining an increasing traction. Standard experimental and data analysis guidelines of ChIP-seq have been proposed and published. Although the antibody quality is a key determining factor for a successful ChIP-seq experiment, the challenge of ChIP-seq undertaking lies at its data analysis. It includes sequencing depth evaluation, quality control of sequencing reads, read mapping, peak calling, peaking annotation, and motif analysis. This chapter focuses on the ChIP-seq data analysis. We will highlight some ChIP-seq applications, summarize typical ChIP-seq data analysis procedures, and demonstrate a practical ChIP-seq data analysis pipeline.

9.2 CHIP-SEQ APPLICATIONS

Application of ChIP-seq is rapidly revolutionizing different areas of science because ChIP-seq is an important experimental technique for studying interactions between specific proteins and DNA in the cell and determining their localization on a specific genomic locus. A variety of phenotypic changes important in normal development and in diseases are temporally and spatially controlled by chromatin-coordinated gene expression. Due to the invaluable ChIP-seq information added to our existing knowledge, considerable progress has been made in our understanding of chromatin structure, nuclear events involved in transcription, transcription regulatory networks, and histone modifications. In this section, Table 9.1 lists several major examples of the important roles that the ChIP-seq approach has played in discovering TF binding sites, the study of TF-mediated different gene regulation, the identification of genome-wide histone marks, and other applications.

9.3 OVERVIEW OF CHIP-SEQ DATA ANALYSIS

Bailey et al. (2013) have published an article entitled "Practical Guidelines for the Comprehensive Analysis of ChIP-seq Data." Interested readers are encouraged to read it in detail. Here we concisely summarize frameworks of these guidelines step by step.

9.3.1 Sequencing Depth

Sequencing depth means the sequencing coverage by sequence reads. The required optimal sequencing depth depends mainly on the size of the genome and the number and size of the binding sites of the protein. For mammalian TFs and chromatin modifications such as enhancer-associated histone marks, which are typically localized at specific, narrow sites and have on the order of thousands of binding sites, 20 million reads may be adequate (4 million reads for worm and fly TFs) (Landt et al. 2012). Proteins with more binding sites (e.g., RNA Pol II) or broader factors, including most histone marks, will require more reads, up to 60 million for mammalian ChIP-seq (Chen et al. 2012).

TABLE 9.1   Representative ChIP-Seq Applications

Usages	Lists	Descriptions	References
Discovering TF-binding sites	1a	The first ChIP-seq experiments to identify 41,582 STAT1-binding regions in IFNγ-HeLa S3 cells.	Robertson et al. (2007)
	1b	ENCODE and modENCODE have performed >1,000 ChIP-seq experiments for >140 TFs and histone modifications in >100 cell types in 4 different organisms.	Landt et al. (2012)
Discovering the molecular mechanisms of TF-mediated gene regulation	2a	Discovered the differential effects of the mutants of lysine 37 and 218/221 of NF-kB p65 in response to IL-1β in HEK 293 cells.	Lu et al. (2013)
	2b	Showed that SUMOylation of the glucocorticoid receptor (GR) modulates the chromatin occupancy of GR on several loci in HEK293 cells.	Paakinaho et al. (2014)
Discovering histone marks	3a	Identified H3K4me3 and H3K27me3 reflecting stem cell state and lineage potential.	Mikkelsen et al. (2007)
	3b	Found H4K5 acetylation and H3S10 phosphorylation associated with active gene transcription.	Park et al. (2013), Tiwari et al. (2011)
Identifying causal regulatory SNPs	4	Detected 4796 enhancer SNPs capable of disrupting enhancer activity upon allelic change in HepG 2 cells.	Huang and Ovcharenko (2014)
Detect disease-relevant epigenomic changes following drug treatment	5	Utilized ChIP-Rx in the discovery of disease-relevant changes in histone modification occupancy.	Orlando et al. (2014)
Decoding the transcriptional regulation of lncRNAs and miRNAs	6	Developed ChIPBase: a database for decoding the transcriptional regulation of lncRNAs and miRNAs.	Yang et al. (2012)

9.3.2 Read Mapping and Quality Metrics

Before mapping the reads to the reference genome, they should be filtered by applying a quality cutoff. These include assessing the quality of the raw reads by Phred quality scores, trimming the end of reads, and examining the library complexity. Library complexity is affected by antibody quality, over-cross-linking, amount of material, sonication, or over-amplification by PCR. Galaxy (galaxy.project.org) contains some tool-boxes for these applications. The quality reads can then be mapped to reference genomes using one of the available mappers such as Bowtie 2 (bowtie-bio.sourceforge.net/bowtie2/), Burrows–Wheeler Aligner (BWA, bio-bwa.sourceforge.net/), Short Oligonucleotide Analysis Package (SOAP, soap.genomics.org.cn/), and Mapping and Assembly with Qualities (MAQ, maq.sourceforge.net/). ChIP-seq data, above 70% uniquely mapped reads, are normal, whereas less than 50% may be cause for concern. A low percentage of uniquely mapped reads often is due to either excessive amplification in the PCR step, inadequate read length, or problems with the sequencing platform. A potential cause of high numbers of *multi-mapping* reads is that the protein binds frequently in regions of repeated DNA. After mapping, the signal-to-noise ratio (SNR) of the ChIP-seq experiment should be assessed, for example, via quality metrics such as strand cross-correlation (Landt et al. 2012) or IP enrichment estimation using the software package CHip-seq ANalytics and Confidence Estimation (CHANCE, github.com/songlab/chance). Very successful ChIP experiments generally have a normalized strand cross-correlation coefficient (NSC) >1.05 and relative strand cross-correlation coefficient (RSC) >0.8. The software CHANCE assesses IP strength by estimating and comparing the IP reads pulled down by the antibody and the background, using a method called *signal extraction* scaling (Diaz et al. 2012).

9.3.3 Peak Calling

A pivotal analysis for ChIP-seq is to predict the regions of the genome where the ChIPed protein is bound by finding regions with significant numbers of mapped reads (peaks). A fine balance between sensitivity and specificity depends on choosing an appropriate peak-calling algorithm and normalization method based on the type of protein ChIPed: point-source factors such as most TFs, broadly enriched factors such as histone marks, and those with both characteristics such as RNA Pol II. SPP and MACS are two peak callers which can analyze all types of ChIPed proteins. SPP (Kharchenko et al. 2008, sites.google.com/a/brown.edu/bioinformatics-in-biomed/spp-r-from-chip-seq) is an R package especially designed for the analysis of ChIP-seq data from Illumina platform. Model-based analysis of ChIP-seq (MACS, liulab.dfci.harvard.edu/MACS/) compares favorably to existing ChIP-seq peak-finding algorithms, is a publicly available open source, and can be used for ChIP-seq with or without control samples. However, it is highly recommended that mapped reads from a control sample be used. Whether comparing one ChIP sample against input DNA (sonicated DNA), *mock* ChIP (non-specific antibody, e.g., IgG) in peak calling, or comparing a ChIP sample against another in differential analysis, there are linear and nonlinear normalization methods available to make the two samples *comparable*. The former includes sequencing depth normalization by a scaling factor, reads per kilobase of sequence range per million mapped reads (RPKM). The latter includes locally weighted regression (LOESS), MAnorm (bcb.dfci.harvard.edu/~gcyuan/MAnorm/). Duplicate reads (same 5′ end) can be removed before peak calling to improve specificity. Paired-end sequencing for ChIP-seq is advocated to improve sensitivity and specificity. A useful approach is to threshold the irreproducible discovery rate (IDR), which, along with motif analysis, can also aid in choosing the best peak-calling algorithm and parameter settings.

9.3.4 Assessment of Reproducibility

To ensure that experimental results are reproducible, it is recommended to perform at least two biological replicates of each ChIP-seq experiment and examine the reproducibility of both the reads and identified peaks. The reproducibility of the reads can be measured by computing the Pearson correlation coefficient (PCC) of the (mapped) read counts at each genomic position. The range of PCC is typically from 0.3 to 0.4 (for unrelated samples) to >0.9 (for replicate samples in high-quality experiments). To measure the reproducibility at the level of peak calling, IDR analysis (Li et al. 2011, www.encodeproject.org/software/idr/) can be applied to the two sets of peaks identified from a pair of replicates. This analysis assesses the rank consistency of identified peaks between replicates and outputs the number of peaks that pass a user-specified reproducibility threshold (e.g., IDR = 0.05). As mentioned above, IDR analysis can also be used for comparing and selecting peak callers and identifying experiments with low quality.

9.3.5 Differential Binding Analysis

Comparative ChIP-seq analysis of an increasing number of protein-bound regions across conditions or tissues is expected with the steady raise of NGS projects. For example, temporal or developmental designs of ChIP-seq experiments can provide different snapshots of a binding signal for the same TF, uncovering stage-specific patterns of gene regulation. Two alternatives have been proposed. The first one—qualitative—implements hypothesis testing on multiple overlapping sets of peaks. The second one—quantitative—proposes the analysis of differential binding between conditions based on the total counts of reads in peak regions or on the read densities, that is, counts of reads overlapping at individual genomic positions. One can use the qualitative approach to get an initial overview of differential binding. However, peaks identified in all conditions will never be declared as differentially bound sites by this approach based just on the positions of the peaks. The quantitative approach works with read counts (e.g., differential binding of TF with ChIP-seq-DBChIP, http://master.bioconductor.org/packages/release/bioc/html/DBChIP.html) computed over peak regions and has higher computational cost, but is recommended as it provides precise statistical assessment of differential binding across conditions (e.g., p-values or q-values linked to read-enrichment fold changes). It is strongly advised to verify that the data fulfill the requirements of the software chosen for the analysis.

9.3.6 Peak Annotation

The aim of the annotation is to associate the ChIP-seq peaks with functionally relevant genomic regions, such as gene promoters, transcription start sites, and intergenic regions. In the first step, one uploads the peaks and reads in an appropriate format (e.g., browser extensible data [BED] or general feature format [GFF] for peaks, WIG or bedGraph for normalized read coverage) to a genome browser, where regions can be manually examined in search for associations with annotated genomic features. The Bioconductor package ChIPpeakAnno (Zhu et al. 2010, bioconductor. org/packages/release/bioc/html/ChIPpeakAnno.html) can perform such *location analyses*, and further correlate them with expression data (e.g., to determine if proximity of a gene to a peak is correlated with its expression) or subjected to a gene ontology analysis (e.g., to determine if the ChIPed protein is involved in particular biological processes).

9.3.7 Motif Analysis

Motif analysis is useful for much more than just identifying the causal DNA-binding motif in TF ChIP-seq peaks. When the motif of the ChIPed protein is already known, motif analysis provides validation of the success of the experiment. Even when the motif is not known beforehand, identifying a centrally located motif in a large fraction of the peaks by motif analysis is indicative of a successful experiment. Motif analysis can also identify the DNA-binding motifs of other proteins that bind in complex or in conjunction with the ChIPed protein, illuminating the mechanisms of transcriptional regulation. Motif analysis is also useful with histone modification ChIP-seq because it can discover unanticipated sequence signals associated with such marks. Table 9.2 lists some publicly available tools for motif analysis.

9.3.8 Perspective

The challenges of ChIP-seq require novel experimental, statistical, and computational solutions. Ongoing advances will allow ChIP-seq to analyze samples containing far fewer cells, perhaps even single cells, greatly expanding its applicability in areas such as embryology and development where large samples are prohibitively expensive or difficult to obtain. No less critical is to trim today's peaks that are much wider than the actual TF binding sites. A promising experimental method for localizing narrow peaks is ChIP-exo that uses bacteriophage λ exonuclease to digest the ends of DNA fragments not bound to protein (Rhee and Pugh 2011). Improving antibody specificity is a long-term endeavor. Another way to eliminate massive amounts of false positive peaks is to limit the regulatory binding sites to nucleosome-depleted regions, which are accessible for regulator binding. Perhaps the most important novel developments are related to the detection and analyses of distal regulatory regions, which are distant in sequence but brought close in 3-D space by DNA bending. To reveal such 3-D mechanisms of transcriptional regulation, two major techniques have emerged: chromatin interaction analysis by paired-end tags (CHIA-PET, Li et al. 2010) and chromosome conformation capture assays such as circular chromosome conformation capture (4C, Van de Werken et al. 2012) or chromosome conformation capture carbon copy (5C, Dostie et al. 2006). It is worth noting that many TFs competitively or cooperatively bind with other TFs, the transcriptional machinery, or cofactors. ChIP-seq will detect *indirect* DNA binding by the protein (via another protein or complex), so predicted sites *not* containing the motif may also be functional. The effects of context-dependent regulatory mechanisms can fundamentally differ from the effects of individual binding events. Binding does not necessarily imply function, so it will remain necessary to use additional information (such as expression or chromatin conformation data) to reliably infer the function of individual binding events.

TABLE 9.2 Software Tools for Motif Analysis of ChIP-Seq Peaks and Their Uses

Category	Software Tool	Web Server	Obtain Peak Regions	Motif Discovery	Motif Comparison	Central Motif Enrichment Analysis	Local Motif Enrichment Analysis	Motif Spacing Analysis	Motif Prediction/ Mapping
Obtaining sequences	Galaxy [50–52][a]	X	X						
	RSAT [53]	X	X						
	UCSC Genome Browser [54]	X	X						
Motif discovery + more	ChIPMunk [55]	X		X					
	CisGenome [56]			X	X				
	CompleteMOTIFS [48]	X		X	X				
	MEME-ChIP [57]	X		X	X	X			
	peak-motifs [58]	X		X	X				X
	Cistrome [49]	X	X	X	X	X	X		X
Motif comparison	STAMP [59]	X			X				
	TOMTOM [60]	X			X				
Motif enrichment/ spacing	CentriMo [61]	X				X	X		
	SpaMo [62]	X						X	
Motif prediction/ mapping	FIMO [63]	X							X
	PATSER [64]	X							X

Source: Bailey, T. et al., *PLoS Comput. Biol.* 9(11), e1003326, 2013.
[a] Those reference numbers are from the above citation.

9.4 STEP-BY-STEP TUTORIAL

The ChIP-seq command pipeline includes read mapping, peak calling, motif detection, and motif region annotation. Here, we use two ChIP-seq data, one from CCCTC-binding factor (CTCF, a zinc finger protein) ChIP-seq experiment (SRR1002555.sra) as case and another from IgG ChIP-seq experiment (SRR1288215.sra) as control in human colon adenocarcinoma cells, which was sequenced using Illumina HiSeq 2000 instrument.

Step 1: To download sra data and convert into FASTQ

```
--------------------------------------------------------------------
# download SRR1002555.sra and SRR1288215.sra data
from NCBI FTP service
$ wget ftp://ftp-trace.ncbi.nlm.nih.gov/sra/sra-
instant/reads/ByExp/sra/SRX%2FSRX360%2FSRX360020/
SRR1002555/SRR1002555.sra
$ wget ftp://ftp-trace.ncbi.nlm.nih.gov/sra/sra-
instant/reads/ByExp/sra/SRX%2FSRX543%2FSRX543697/
SRR1288215/SRR1288215.sra
# covert sra format into fastq format
$ fastq-dump SRR1002555.sra
$ fastq-dump SRR1288215.sra
# when it is finished, you can check all files:
$ ls -l
# SRR1002555.fastq and SRR1288215.fastq will be
produced.
--------------------------------------------------------------------
```

Step 2: To prepare human genome data and annotation files

```
--------------------------------------------------------------------
# downloading human hg19 genome from illumina
iGenomes and gene annotation table with genome
background annotations from CEAS
$ wget ftp://igenome:G3nom3s4u@ussd-ftp.illumina.com/
Homo_sapiens/UCSC/hg19/Homo_sapiens_UCSC_hg19.tar.gz
$ wget http://liulab.dfci.harvard.edu/CEAS/src/hg19.
refGene.gz
# gunzip .gz files
$ gunzip *.gz
# linking human genome and bowtie index into current
working direction
```

```
$ In -s /homo.sapiens/UCSC/hg19/Sequence/
WholeGenomeFasta/genome.fa
$ In -s /homo.sapiens/UCSC/hg19/Sequence/
BowtieIndex/genome.1.ebwt
$ In -s /homo.sapiens/UCSC/hg19/Sequence/
BowtieIndex/genome.2.ebwt
$ In -s /homo.sapiens/UCSC/hg19/Sequence/
BowtieIndex/genome.3.ebwt
$ In -s /homo.sapiens/UCSC/hg19/Sequence/
BowtieIndex/genome.4.ebwt
$ In -s /homo.sapiens/UCSC/hg19/Sequence/
BowtieIndex/genome.rev.1.ebwt
$ In -s /homo.sapiens/UCSC/hg19/Sequence/
BowtieIndex/genome.rev.2.ebwt
# when it is finished, you can check all files:
$ ls -l
# genome.fa, genome.1.ebwt, genome.2.ebwt,
genome.3.ebwt, genome.4.ebwt, genome.rev.1.ebwt,
genome.rev.2.ebwt and hg19.refGene will be produced.
```

Step 3: Mapping the reads with Bowtie

For ChIP-seq data, the currently common programs are BWA and Bowtie. Here, we will use Bowtie as example. The parameter genome is human hg19 genome index; -q query input files are FASTQ; -v 2 will allow two mismatches in the read, when aligning the read to the genome sequence; -m 1 will exclude the reads that do not map uniquely to the genome; -S will output the result in SAM format.

```
$ bowtie genome -q SRR1002555.fastq -v 2 -m 1 -S >
CTCF.sam
$ bowtie genome -q SRR1288215.fastq  -v 2 -m 1 -S >
lgG.sam
# when it is finished, you can check all file:
$ ls -l
# CTCF.sam and lgG.sam will be produced.
```

Step 4: Peak calling with MACS

Macs callpeak is used to call peaks where studied factor is bound from alignment results. The output files of bowtie (CTCF.sam and lgG.sam) will be the input of macs. The parameters -t and -c are used to define the names of case (CTCF.sam) and control (lgG.sam); -f SAM query input files are SAM; --gsize 'hs' defines human effective genome size; --name "CTCF" will be used to generate output file names; --bw 400 is the band width for picking regions to compute fragment size; --bdg will output a file in bedGraph format to visualize the peak profiles in a genome browser. The output files CTCF_peaks.xls and CTCF_peaks. narrowPeak will give us details about peak region.

```
$ macs callpeak -t CTCF.sam -c lgG.sam -f SAM
--gsize 'hs' --name "CTCF" --bw 400 --bdg
# when it is finished, you can check all file:
$ ls -l
# CTCF_treat_pileup.bdg, CTCF_summits.bed, CTCF_
peaks.xls, CTCF_peaks.narrowPeak and CTCF_control_
lambda.bdg will be produced.
```

Step 5: Motif analysis

Multiple EM for Motif Elicitation-ChIP (MEME-ChIP) will be used to discover DNA-binging motifs on a set of DNA sequences from peak regions. Before running MEME-ChIP, we use bedtools getfasta to prepare binding region sequences. The output file (peak.fa) will be as input of MEME-ChIP. The MEME-ChIP parameter -meme-p defines parallel processors in the cluster; -oc defines the output to the specified directory, overwriting if the directory exists. The output file index.html gives us the significant motifs (E-value ≤ 0.05) found by the programs MEME, Discriminative Regular Expression Motif Elicitation (DREME), and CentriMo (maximum central enrichment) and running status.

```
# preparing sequences corresponding the peaks
$ bedtools getfasta -fi genome.fa -bed CTCF_peaks.
narrowPeak -fo peak.fa
# running meme-chip for CTCF motif
$ meme-chip -meme-p 6 -oc CTCF-meme-out peak.fa
# when it is finished, you can check all file:
$ ls -l
# CTCF-meme-out directory will be produced, which
contain all motifs detail.
```

Step 6: ChIP region annotation

CEAS (Cis-regulatory Element Annotation System) provides statistics on ChIP enrichment at important genome features such as specific chromosome, promoters, gene bodies, or exons and infers genes most likely to be regulated by a binding factor. The input files are gene annotation table file (hg19.refGene) and BED file with ChIP regions (CTCF.bed). Output file CTCF_ceas.pdf will print genome features distribution of ChIP regions; CTCF_ceas.xls will tell the details of genome features distribution.

```
# preparing bed file with ChIP regions
$ cut CTCF_peaks.narrowPeak -f 1,2,3 > CTCF.bed
# running ceas using default mode
$ ceas --name=CTCF_ceas -g hg19.refGene -b CTCF.bed
# when it is finished, you can check all file:
$ ls -l
#  CTCF_ceas.pdf, CTCF_ceas.R and CTCF_ceas.xls will
be produced.
```

BIBLIOGRAPHY

Bailey T, Krajewski P, Ladunga I, Lefebvre C, Li Q, et al. Practical guidelines for the comprehensive analysis of ChIP-seq data. *PLoS Comput Biol.* 2013; 9(11):e1003326. doi:10.1371/journal.pcbi.1003326.

Barski A, Zhao K. Genomic location analysis by ChIP-Seq. *J Cell Biochem.* 2009; 107(1):11–18.

Furey TS. ChIP-seq and beyond: New and improved methodologies to detect and characterize protein-DNA interactions. *Nat Rev Genet.* 2012;13(12):840–52.

http://www.illumina.com/Documents/products/datasheets/datasheet_chip_sequence.pdf.

Johnson DS, Mortazavi A, Myers RM, Wold B. Genome-wide mapping of in vivo protein-DNA interactions. *Science.* 2007; 316(5830):1497–502.

Kim H, Kim J, Selby H, Gao D, Tong T, Phang TL, Tan AC. A short survey of computational analysis methods in analysing ChIP-seq data. *Hum Genomics.* 2011; 5(2):117–23.

Landt SG, Marinov GK, Kundaje A, Kheradpour P, Pauli F, et al. ChIP-seq guidelines and practices of the ENCODE and modENCODE consortia. *Genome Res.* 2012; 22(9):1813–31.

Mundade R, Ozer HG, Wei H, Prabhu L, Lu T. Role of ChIP-seq in the discovery of transcription factor binding sites, differential gene regulation mechanism, epigenetic marks and beyond. *Cell Cycle.* 2014; 13(18):2847–52.

Park PJ. ChIP-seq: Advantages and challenges of a maturing technology. *Nat Rev Genet.* 2009; 10(10):669–80.

Yang JH, Li JH, Jiang S, Zhou H, Qu LH. ChIPBase: A database for decoding the transcriptional regulation of long non-coding RNA and microRNA genes from ChIP-Seq data. *Nucleic Acids Res.* 2013; 41:D177–87.

III

Integrative and Comprehensive Big Data Analysis

## CHAPTER 10

Integrating Omics Data in Big Data Analysis

Li Qin Zhang, Daniel P. Heruth, and Shui Qing Ye

CONTENTS

10.1 INTRODUCTION

The relatively newly coined word *omics* refers to a field of study in biology ending in *-omics*, such as genomics, transcriptomics, proteomics, or metabolomics. The related suffix -ome is used to address the objects of study of such fields, such as the genome, transcriptome, proteome, or metabolome, respectively. Omics aims at the collective characterization and quantification of pools of biological molecules that translate into the structure, function, and dynamics of an organism or organisms. For example, genomics is to sequence, assemble, and analyze the structure and function of the complete set of DNA within an organism. Omics becomes a buzz word; it is increasingly added as a suffix to more fields to indicate the *totality* of those fields to be investigated such as connectomics to study the totality of neural connections in the brain; interactomics to engage in analyses of all gene–gene, protein–protein, or protein–RNA interactions within a system; and lipidomics to study the entire complement of cellular lipids within a cell or tissue or organism. Now, another term *panomics* has been dubbed to refer to all omics including genomics, proteomics, metabolomics, transcriptomics, and so forth, or the integration of their combined use.

The advent of next-generation DNA sequencing (NGS) technology has fueled the generation of omics data since 2005. Two hallmarks of NGS technology that distinguish it from the first-generation DNA sequencing technology are faster speed and lower cost. At least three technical advances have made the development of NGS technology possible or practical to realize. First, general progress in technology across disparate fields, including microscopy, surface chemistry, nucleotide biochemistry, polymerase engineering, computation, data storage, and others, has provided building blocks or foundations for the production of NGS platforms. Second, the availability of whole-genome assemblies for *Homo sapiens* and other model organisms provides references against which short reads, typical of most NGS platforms, can be mapped or aligned. Third, a growing variety of molecular methods has been developed, whereby a broad range of biological phenomena can be assessed to elucidate the role and functions of any gene in health and disease, thus increasing demand of gene sequence information by high-throughput DNA sequencing (e.g., genetic variation, RNA expression, protein–DNA interactions, and chromosome conformation). Over the past 10 years, several platforms of NGS technologies, as detained in previous chapters of this book, have emerged as new and more powerful strategies for DNA sequencing, replacing the first-generation DNA sequencing technology based on the Sanger method as a preferred technology for high-throughput DNA sequencing tasks. Besides directly generating omics data such as genomics, epigenomics, microbiomics, and transcriptomics, NGS has also fueled or driven the development of other technologies to facilitate the generation of other omics data such as interactomics, metabolomics, and proteomics.

Understanding the genetic basis of complex traits has been an ongoing quest for many researchers. The availability of rich omics data has made possible to derive global molecular insights into health and disease. Historically and currently, many investigators have ventured to probe each type of omics data independently to look for relationships with biological processes. Using these methods, some of the pieces of the puzzle of complex-trait genetic architecture and basic biological pathways have been successfully untangled. However, much of the genetic etiology of complex traits and biological networks remains unexplained, which could be partly due to the focus on restrictive single-data-type study designs. Recognizing this limitation, integrated omics data analyses have been used increasingly. This integrated omics approach can achieve a more thorough and informative interrogation of genotype–phenotype associations than an analysis that uses only a single data type. Combining multiple data types can compensate for missing or unreliable information in any single data type, and multiple sources of evidence pointing to the same gene or pathway are less likely to lead to false positives. Importantly, the complete biological model is only likely to be discovered if the different levels of omics data are considered in an analysis. In this chapter, we will highlight some successful applications of integrated omics data analysis, synopsize most important strategies in integrated omics data analysis, and demonstrate one special example of such integrated omics data analysis.

10.2 APPLICATIONS OF INTEGRATED OMICS DATA ANALYSIS

The realization that performing all analyses from a single source or within one data type has limitations has spurred applications of integrated omics data analyses. Although these systematic approaches are still in their infancy, they have shown promise to perform powerful integrative analyses, some of which may lay solid foundations to become gold standard methods of future integrated omics data analyses down the road. Chen et al. (2012) reported the first integrative personal omics profile (iPOP), an analysis that combines genomic, transcriptomic, proteomic, metabolomic, and autoantibody profiles from a single individual over a 14-month period. Their iPOP analysis revealed various medical risks, including type 2 diabetes. It also uncovered extensive, dynamic changes in diverse molecular components and biological pathways across healthy and diseased conditions. The integrated omics data analysis also revealed extensive heteroallelic changes during healthy and diseased states and an unexpected RNA editing mechanism. This *trail blazing* study demonstrates that longitudinal iPOP can be used to interpret healthy and diseased states by connecting genomic information with additional dynamic omics activity. Here Table 10.1 lists some representative integrating omics data analyses examples.

TABLE 10.1   Representative Application of Integrated Omics Data Analysis

#	Application	Software	Website	References
1	Meta-analysis of gene expression data	INMEX	inmex.ca/INMEX/	Xia et al. (2013)
2	eQTL	Matrix eQTL	www.bios.unc.edu/ research/genomic_ software/Matrix_eQTL/	Shabalin et al. (2012)
3	A searchable human eQTL database	seeQTL	http://www.bios.unc.edu/ research/genomic_ software/seeQTL/	Xia et al. (2012)
4	Methylation QTL	Scan database	www.scandb.org/	Zhang et al. (2015)
5	Protein QTL	pQTL	eqtl.uchicago.edu/cgi-bin/ gbrowse/eqtl	Hause et al. (2014)
6	Allele-specific expression	AlleleSeq	alleleseq.gersteinlab.org/	Rozowsky et al. (2011)
7	Functional annotation of SNVs	Annovar Regulome DB	www. openbioinformatics. org/annovar/ www.regulomedb.org	Wang et al. (2010), Boyle et al. (2012)
8	Concatenational integration	Athena WinBUGS Glmpath	ritchielab.psu.edu/ ritchielab/software www.mrc-bsu.cam.ac.uk/ software/ cran.r-project.org/web/ packages/glmpath/index. html	Holzinger et al. (2014), Lunn et al. (2000), Park et al. (2013)
9	Transformational integration	SKmsmo Gbsll	imagine.enpc.fr/~obozinsg/ SKMsmo.tar mammoth.bcm.tmc.edu/ papers/lisewski2007.gz	Lanckriet et al. (2004), Kim et al. (2012)
10	Model-based integration	Ipred Weka3	cran.r-project.org/web/ packages/ipred/index.html www.cs.waikato.ac.nz/ml/ weka/	Peters et al. (2015), Akavia et al. (2010)

10.3 OVERVIEW OF INTEGRATING OMICS DATA ANALYSIS STRATEGIES

Ritchie et al. (2015) have recently published an elegant review on "Methods of integrating data to uncover genotype–phenotype interactions." Interested readers are encouraged to refer to this review for details. When combining or integrating omics data, there are unique challenges for individual data types, and it is important to consider these before implementing meta-, multi-staged, or meta-dimensional analyses; these include data quality, data scale or dimensionality, and potential confounding of the data (see below). If these issues are not dealt with for each individual data type, then they could cause problems when the data types are integrated. Due to the space limitation, this section won't cover the quality control, data reduction, and confounding factor adjust of each individual data type.

10.3.1 Meta-Analysis

Gene V. Glass, an American statistician, coined the term *meta-analysis* and illustrated its first use in his presidential address to the American Educational Research Association in San Francisco in April 1976. Meta-analysis comprises statistical methods for contrasting and combining results from different studies in the hope of identifying patterns among study results, sources of disagreement among those results, or other interesting relationships that may come to light in the context of multiple studies. Meta-analysis can be thought of as *conducting research about previous research* or *the analysis of analyses.* The motivation of a meta-analysis is to aggregate information in order to achieve a higher statistical power for the measure of interest, as opposed to a less precise measure derived from a single study. Usually, five steps are involved in a meta-analysis: (1) formulation of the problem; (2) search for the literature; (3) selection of studies; (4) decide which dependent variables or summary measures are allowed; and (5) selection and application of relevant statistic methods to analyze the metadata. Xia et al. (2013) introduced the integrative meta-analysis of expression data (INMEX), a user-friendly web-based tool (inmex.ca/INMEX/) designed to support meta-analysis of multiple gene-expression data sets, as well as to enable integration of data sets from gene expression and metabolomics experiments. INMEX contains three functional modules. The data preparation module supports flexible data processing, annotation, and visualization of individual data sets. The statistical analysis module allows researchers to combine multiple data sets based on p-values, effect sizes, rank orders, and other features. The significant genes can be examined in functional analysis module for enriched gene ontology terms or Kyoto Encyclopedia of Genes and Genomes (KEGG) pathways, or expression profile visualization.

10.3.2 Multi-Staged Analysis

Multi-staged analysis, as its name suggests, aims to divide data analysis into multiple steps, and signals are enriched with each step of the analysis. The main objective of the multi-staged approach is to divide the analysis into multiple steps to find associations first between the different data types, then subsequently between the data types and the trait or phenotype of interest. Multi-staged analysis is based on the assumption that variation is hierarchical, such that variation in DNA leads to variation in RNA and so on in a linear manner, resulting in a phenotype. There have been three types of analysis methods in this category: *genomic variation analysis approaches*, *allele-specific expression approaches*, and *domain knowledge-guided approaches*.

In *genomic variation analysis approaches*, the rationale is that genetic variations are the foundation of all other molecular variations. This approach generally consists of three-stage analyses. Stage 1 is to associate SNPs with the phenotype and filter them based on a genome-wide significance threshold. Stage 2 is to test significant SNPs from stage 1 for association with another level of omic data. For example, one option is to look for the association of SNPs with gene expression levels, that is, expression quantitative trait loci (eQTLs), and alternatively, to examine SNPs associated with DNA methylation levels (methylationQTL), metabolite levels (metaboliteQTL), protein levels (pQTLs), or other molecular traits such as long non-coding RNA and miRNA. Illustrating this approach, Huang et al. (2007) first described an integrative analysis to identify DNA variants and gene expressions associated with chemotherapeutic drug (etoposide)-induced cytotoxicity. One of challenges for this approach arises when a relatively arbitrary threshold, generally a *p*-value, is used to identify the significant associations for further analyses. As the *p*-value threshold also needs to be adjusted for the number of tests being carried out to combat multiple testing problems, there is likely to be a large number of false-negative SNPs, eQTLs, mQTLs, and pQTLs being filtered out.

In *allele-specific expression (ASE) approaches*, the rationale is that in diploid organisms, genetic variation occurs at one of the two alleles, which alters the regulation of gene expression, leads to allele-specific expression in some genes, and hence contributes to phenotypic variation. ASE variants are associated with *cis*-element variations and epigenetic modifications. The first step of ASE approaches is to distinguish the gene product of one parental allele from the product of the other parental allele. Step 2 is to associate the allele with gene expression (eQTLs) or methylation (mQTLs) or others to compare the two alleles. Step 3 is to test the resulting alleles for correlation with a phenotype or an outcome of interest. ASE has been applied to identify functional variations from hundreds of multi-ethnic individuals from the 1000 Genome Project (Lappalainen et al. 2013), to map allele-specific protein–DNA interactions in human cells (Maynard et al. 2008), and to explore allele-specific chromatin state (Kasowski et al. 2013) and histone modification (McVicker et al. 2013). The analysis of allele-specific transcription offers the opportunity to define the identity and mechanism of action of cis-acting regulatory genetic variants that modulate transcription on a given chromosome to shed new insights into disease risk.

In *domain knowledge-guided approaches,* the genomic regions of interest are inputs to be used to determine whether the regions are within pathways and/or overlapping with functional units, such as transcription factor binding, hypermethylated or hypomethylated regions, DNase sensitivity, and regulatory motifs. In this approach, step 1 is to take a collection of genotyped SNPs and annotate them with domain knowledge from multiple public database resources. Step 2 is to associate functional annotated SNPs with other omic data. Step 3 is to evaluate positive targets selected from step 2 for correlation with a phenotype or an outcome of interest. Many available public knowledge databases or resources such as ENCyclopedia Of DNA Elements (ENCODE, www.encodeproject.org) and the Kyoto Encyclopedia of Genes and Genomes (KEGG, www.genome.jp/kegg/) have made this approach feasible and practical. This approach adds information from diverse data sets that can substantially increase our knowledge of our data; however, we are also limited and biased by current knowledge.

10.3.3 Meta-Dimensional Analysis

The rationale behind meta-dimensional analysis is that it is the combination of variation across all possible omic levels in concert that leads to phenotype. Meta-dimensional analysis combines multiple data types in a simultaneous analysis and is broadly categorized into three approaches: *concatenation-based integration, transformation-based integration, and model-based integration.*

In *concatenation-based integration,* multiple data matrices for each sample are combined into one large input matrix before a model is constructed as shown in Figure 10.1a. The main advantage of this approach is that it can factor in interactions between different types of genomic data. This approach has been used to integrate SNP and gene expression data to predict high-density lipoprotein cholesterol levels (Holzinger et al. 2013) and to identify interactions between copy number alteration, methylation, miRNA, and gene expression data associated with cancer clinical outcomes (Kim et al. 2013). Another advantage of concatenation-based integration is that, after it is determined how to combine the variables into one matrix, it is relatively easy to use any statistical method for continuous and categorical data for analysis. For example, Fridley et al. (2012) modeled the joint relationship of mRNA gene expression and SNP genotypes using a Bayesian integrative model to predict a quantitative phenotype such as drug gemcitabine cytotoxicity. Mankoo et al. (2011) predicted time to recurrence and survival in ovarian cancer using copy number alteration, methylation, miRNA, and gene expression data using a multivariate Cox LASSO (least absolute shrinkage and selection operator) model.

The challenge with concatenation-based integration is identifying the best approach for combining multiple matrices that include data from different scales in a meaningful way without biases driven by data type. In addition, this form of data integration can inflate high dimensionality for the data, with the number of samples being smaller than the number of measurements for each sample (Clarke et al. 2008). Data reduction strategies may be needed to limit the number of variables to make this analysis possible.

In *transformation-based integration,* multiple individual data type is individually transformed into its corresponding intermediate form such as graph or kernel matrix before they are merged and then modeled (Figure 10.1b). A graph is a natural method for analyzing relationships between samples, as the nodes depict individual samples and the edges represent their possible relationships. Kernel matrix is a symmetrical and positive semi-definite matrix that represents the relative positions of all samples conducted by valid kernel functions. The transformation-based integration approach can preserve data-type-specific properties when each data type is appropriately transformed

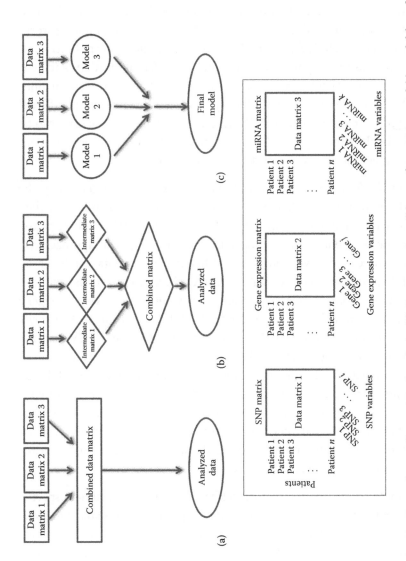

FIGURE 10.1 Categorization of meta-dimensional analysis. (a) Concatenation-, (b) transformation-, and (c) model-based integrations. (Modified from Ritchie, M.D. et al. *Nat. Rev. Genet.* 16, 85–97, 2015.)

into an intermediate representation. It can be used to integrate many types of data with different data measurement scales as long as the data contain a unifying feature. Kernel-based integration has been used for protein function prediction with multiple types of heterogeneous data (Lanckriet et al. 2004; Borgwardt et al. 2005). Graph-based integration has been used to predict protein function with multiple networks (Suda et al. 2005; Shin et al. 2007) and to predict cancer clinical outcomes using copy number alteration, methylation, miRNA, and gene expression (Kim et al. 2012). The disadvantage of transformation-based integration is that identifying interactions between different types of data (such as a SNP and gene expression interaction) can be difficult if the separate transformation of the original feature space changes the ability to detect the interaction effect.

In *model-based integration*, multiple models are generated using the different types of data as training sets, and a final model is then generated from the multiple models created during the training phase, preserving data-specific properties (Figure 10.1c). This approach can combine predictive models from different types of data. Model-based integration has been performed with ATHENA to look for associations between copy number alterations, methylation, microRNA, and gene expression with ovarian cancer survival (Kim et al. 2013). A *majority voting* approach was used to predict drug resistance of HIV protease mutants (Drăghici et al. 2003). Ensemble classifiers have been used to predict protein-fold recognition (Shen et al. 2006). Network-based approaches such as Bayesian network have been employed to construct probabilistic causal networks (Akavia et al. 2010). In each of these model-based integration examples, a model is built on each data type individually, and the models are then combined in some meaningful way to detect integrative models.

It should be pointed out that model-based integration requires a specific hypothesis and analysis for each data type, and a mechanism to combine the resulting models in a meaningful way. This approach may miss some of the interactions between different data types. Therefore, model-based integration is particularly suitable if each genomic data type is extremely heterogeneous, such that combining the data matrix (concatenation-based integration) or performing data transformation to a common intermediate format (transformation-based integration) is not possible.

10.3.4 Caveats for Integrating Omics Data Analysis

It is critical that the assumptions of the model, limitations of the analysis, and caution about inference and interpretation be taken into consideration for a successful multi-omic study.

The *gold standard* in human genetics is to look for replication of results using independent data, and seeking replication of multi-omic models is one way to identify robust predictive models to avoid or minimize false discoveries. Functional validation is a viable alternative to replication. For example, basic experimental bench science can be used to provide validation for statistical models. Another validation approach is the use of text mining to find literature that supports or refutes the original findings. *In silico* modeling is an additional approach that can be useful.

As more data are generated across multiple data types and multiple tissues, novel explorations will further our understanding of important biological processes and enable more comprehensive systems genomic strategies. It is through collaboration among statisticians, mathematicians, computer scientists, bioinformaticians, and biologists that the continued development of meta-dimensional analysis methods will lead to a better understanding of complex-trait architecture and generate new knowledge about human disease and biology.

10.4 STEP-BY-STEP TUTORIAL

Here we demonstrate the iClusterPlus program developed by Mo and Shen (2013, http://www.bioconductor.org/packages/release/bioc/html/iClusterPlus.html) to perform an integrative clustering analysis of somatic mutation, DNA copy number, and gene expression data from a glioblastoma data set in The Cancer Genome Atlas (TCGA). The iClusterPlus is an R package for integrative clustering of multiple genomic data sets tool using a joint latent variable model. It can extract useful information from multiple omic data (genome data, transcriptome data, epigenome data, proteome data, and phenotype data) to study biological meaning, disease biomarker, and driver genes. Before following this tutorial, readers need to make sure that R package in the computer cluster is available at his or her disposal. If not, readers can follow the instruction from the nearest mirror site (http://cran.r-project.org/mirrors.html) to download and install it. Also, readers can follow the instruction (http://www.bioconductor.org/packages/release/bioc/html/iClusterPlus.html) to install iClusterPlus program. The detailed explanation of the following tutorial steps can be found in the manual (http://www.bioconductor.org/packages/release/bioc/vignettes/iClusterPlus/inst/doc/iManual.pdf).

Step 1: Install iClusterPlus and other package

```
> source ("http://bioconductor.org/biocLite.R")
> biocLite ("iClusterPlus")
> biocLite ("GenomicRanges ")
> biocLite ("gplots")
> biocLite ("lattice")
```

Step 2: Load different package

```
# load iClusterPlus, GenomicRanges, gplots and
lattice package and gbm data package (TCGA
glioblastoma data set)
> library (iClusterPlus)
> library (GenomicRanges)
> library (gplots)
> library (lattice)
> data (gbm)
```

Step 3: Pre-process data

```
# prepare mutation data set, pick up mutations of
which average frequency are bigger than 2%
> mut.rate=apply (gbm.mut, 2, mean)
> gbm.mut2 = gbm.mut [,which (mut.rate>0.02)]
# load human genome variants of the NCBI 36 (hg18)
assembly package
> data (variation.hg18.v10.nov.2010)
# reduce the GBM copy number regions to 5K by
removing the redundant regions using
function CNregions
> gbm.cn=CNregions (seg=gbm.seg, epsilon=0, adaptive=
FALSE, rmCNV=TRUE, cnv=variation.hg18.v10.nov.2010
[,3:5], frac.overlap=0.5,
rmSmallseg=TRUE, nProbes=5)
> gbm.cn=gbm.cn [order (rownames (gbm.cn)),]
```

Step 4: Integrative clustering analysis

```
--------------------------------------------------------------------------------
# use iClusterPlus to integrate GBM mutation data
(gbm.mut2), copy number variation data (gbm.cn), and
gene expression data (gbm.exp). The parameters dt1,
dt2, dt3 require input data matrix; type means
distribution of your data; lambda means vector of
lasso penalty terms; K means number of eigen
features, the number of cluster is K+1; maxiter
means maximum iteration for the EM algorithm.
>fit.single=iClusterPlus(dt1=gbm.mut2,dt2=gbm.cn,
dt3=gbm.exp, type=c("binomial","gaussian",
"gaussian"),lambda=c(0.04,0.05,0.05),K=5,maxiter=10)
> fit.single$alpha
# alpha is intercept parameter of each marker, region
and gene.
> fit.single$beta
# beta is information parameter of each marker, region
and gene.
> fit.single$clusters
# clusters is sample cluster assignment.
> fit.single$centers
# centers is cluster center.
> fit.single$meanZ
# meanZ is latent variable.
> fit.single$BIC
# BIC is Bayesian information criterion.
--------------------------------------------------------------------------------
```

Step 5: Generate heatmap

```
--------------------------------------------------------------------------------
# set maximum and minimum value for copy number
variation and gene expression
> cn.image=gbm.cn
> cn.image[cn.image>1.5]=1.5
> cn.image[cn.image< -1.5]= -1.5
> exp.image=gbm.exp
> exp.image[exp.image>2.5]=2.5
> exp.image[exp.image< -2.5]= -2.5
# set heatmap color for SNP, copy number variation
and gene expression data
```

```
> bw.col = colorpanel(2,low="white",high="black")
> col.scheme = alist()
> col.scheme[[1]] = bw.col
> col.scheme[[2]] = bluered(256)
> col.scheme[[3]] = bluered(256)
# generate heatmap for 6 clusters of 3 different
data sets
> pdf("heatmap.pdf",height=6,width=6)
> plotHeatmap(fit=fit.single,datasets=list(gbm.
mut2,cn.image,exp.image),type=c("binomial","gaussian",
"gaussian"), col.scheme = col.scheme,
row.order=c(T,T,T),chr=chr,plot.chr=c(F,F,F),sparse=c
(T,T,T),cap=c(F,F,F))
> dev.off()
# if you follow the tutorial correctly, the plot as
in Figure 10.2 should appear in your folder.
```

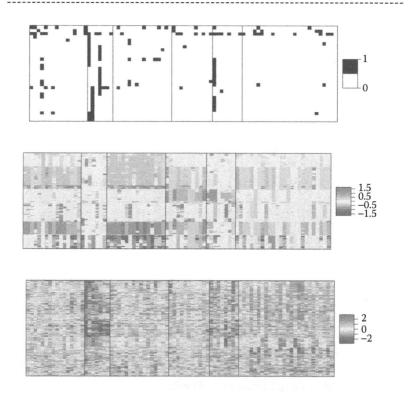

FIGURE 10.2 Heatmap of mutation (top panel), DNA copy number (middle panel), and mRNA expression (bottom panel) for the three-cluster solution. Rows are genomic features, and columns are samples.

Step 6: Feature selection

```
# select the top features based on lasso coefficient
estimates for the 6-cluster solution
> features = alist()
> features[[1]] = colnames(gbm.mut2)
> features[[2]] = colnames(gbm.cn)
> features[[3]] = colnames(gbm.exp)
> sigfeatures=alist()
> for(i in 1:3){
rowsum=apply(abs(fit.single$beta[[i]]),1, sum)
upper=quantile(rowsum,prob=0.75)
sigfeatures[[i]]=(features[[i]])
[which(rowsum>upper)]
}
> names(sigfeatures)=c("mutation","copy number",
"expression")
# top mutant feature markers
> head(sigfeatures[[1]])
If you follow the tutorial correctly, the following
result should appear:
[1] "A2M"        "ADAMTSL3" "BCL11A"     "BRCA2"
"CDKN2A"     "CENTG1"

# top copy number variation feature regions
> head(sigfeatures[[3]])
If you follow the tutorial correctly, the following
result should appear:
[1] "chr1.201577706-201636128"
"chr1.201636128-202299299"
[3] "chr1.202299299-202358378"
"chr1.202358378-202399046"
[5] "chr1.202399046-202415607"
"chr1.202415607-202612588"

# top expression feature genes
> head(sigfeatures[[2]])
If you follow the tutorial correctly, the following
result should appear:
[1] "FSTL1"     "BBOX1"     "CXCR4"     "MMP7"
"ZEB1"       "SERPINF1"
```

BIBLIOGRAPHY

1. Ritchie MD, Holzinger ER, Li R, Pendergrass SA, Kim D. Methods of integrating data to uncover genotype-phenotype interactions. *Nat Rev Genet.* 2015; 16(2):85–97.
2. Cheranova D, Zhang LQ, Heruth D, Ye SQ. Chapter 6: Application of next-generation DNA sequencing in medical discovery. In *Bioinformatics: Genome Bioinformatics and Computational Biology.* 1st ed., pp. 123–136, ed. Tuteja R, Nova Science Publishers, Hauppauge, NY, 2012.
3. Hawkins RD, Hon GC, Ren B. Next-generation genomics: An integrative approach. *Nat. Rev. Genet.* 2010; 11:476–486.
4. Holzinger ER, Ritchie MD. Integrating heterogeneous high-throughput data for meta-dimensional pharmacogenomics and disease-related studies. *Pharmacogenomics* 2012; 13:213–222.
5. Chen R, Mias GI, Li-Pook-Than J et al. Personal omics profiling reveals dynamic molecular and medical phenotypes. *Cell* 2012; 148(6):1293–1307.
6. Gehlenborg N, O'Donoghue SI, Baliga NS et al. Visualization of omics data for systems biology. *Nat Methods.* 2010; 7(3 Suppl):S56–S68.
7. Vogelstein B, Papadopoulos N, Velculescu VE, Zhou S, Diaz LA Jr, Kinzler KW. Cancer genome landscapes. *Science.* 2013; 339(6127):1546–1558.
8. Kodama K, Tojjar D, Yamada S, Toda K, Patel CJ, Butte AJ. Ethnic differences in the relationship between insulin sensitivity and insulin response: A systematic review and meta-analysis. *Diabetes Care.* 2013; 36(6):1789–1996.
9. Chervitz SA, Deutsch EW, Field D et al. Data standards for omics data: The basis of data sharing and reuse. *Methods Mol Biol.* 2011; 719:31–69.
10. Huber W, Carey VJ, Gentleman R et al. Orchestrating high-throughput genomic analysis with Bioconductor. *Nat Methods.* 2015; 12(2):115–121.
11. Mo Q, Shen R. iClusterPlus: Integrative clustering of multi-type genomic data. R package version 1.4.0. 2013, http://www.bioconductor.org/packages/release/bioc/html/iClusterPlus.html.

CHAPTER **11**

Pharmacogenetics and Genomics

Andrea Gaedigk, Katrin Sangkuhl,
and Larisa H. Cavallari

CONTENTS

11.1 INTRODUCTION

The term *pharmacogenetics* was first coined in 1959 by Vogel after Motulsky published his seminal work describing observations that mutations in drug-metabolizing enzymes are associated with a toxic response to drugs. Today, this term is used to describe genetic variation in genes contributing to interindividual drug response and adverse drug events. Genes involved in drug absorption, distribution, metabolism, and elimination, also known as *ADME genes* (http://pharmaadme.org/), include many phase I drug metabolizing enzymes of the cytochrome P450 superfamily such as *CYP2C9*, *CYP2C19*, and *CYP2D6*; phase II drug metabolizing enzymes such as UDP glucuronosyltransferases, glutathione transferases, and thiopurine *S*-methyltransferase; and drug transporters of the ATP-binding cassette ABC and solute carrier SLC families. Detailed summaries of the most prominent pharmacogenes, clinical perspectives, and applications as well as other topics related to pharmacogenetics and genomics have been extensively covered [1,2].

The term *pharmacogenomics* emerged in the late 1990 and early 2000s after technological advances in genome analysis and bioinformatics allowed studies to expand from single/few gene approaches to study many genes (the genome) and pathways to more systematically explore the role of genetic variation on drug response, therapeutic failure, and adverse events. While pharmacogenetics and pharmacogenomics are often used interchangeably, the latter more accurately reflects recent efforts to integrate the genome, epigenome, transcriptome, proteome, and metabolome into a unifying discipline [3]. Likewise, the term *phenotype* is typically used within the context of pharmacogenetics to describe the capacity of an individual to metabolize a drug of interest. For example, the cough suppressant dextromethorphan is administered and metabolites measured in the urine (Table 11.1), and the urinary ratio of dextromethorphan/dextrorphan is then employed as a surrogate measure of *CYP2D6* activity. It is, however, increasingly appreciated that a person's phenotype is rather complex and likely is the culmination of variations residing in all the aforementioned omic levels combined. Hence, the term *phenome* more precisely describes a disease state (cancer, diabetes, and autism) or a drug response trait (responder and nonresponder) [3].

As summarized in Section 11.2, pharmacogenetics and genomics, from here on referred to as PGx, utilize a plethora of methods and approaches to study relationships between genes/genome and phenotype/phenome. Among those are genome-wide association studies (GWAS) to explore relationships between genetic variation and pharmacokinetic and pharmacodynamic effects as well as adverse events [4–6] (examples are provided in Section 11.2). Often, sequence variation(s) detected in GWAS studies only reveal associations that partially explain the observed variability. Ritchie and coauthors review current methods and explore emerging approaches to integrate big omics data to reveal relationships between genomic variation and phenome. They also argue that there is a need for even more advanced analysis strategies to utilize the high-throughput omic data to discover not only true associations, but also associations that are currently missed [3]. In addition to the host omic-composition, there is a growing body of evidence suggesting that the gut microbiome not only affects host physiology and health (see Chapter 9), but is also contributing to interindividual drug

TABLE 11.1   Definitions of Terms Central to PGx

#	Method/Strategy	Description	References
1	Phenotype	The terms *phenotype* and *phenotyping* are often used in PGx to describe a person's metabolic capacity to metabolize drugs of interest. To that end, drugs and/or metabolites are measured in urine and metabolic ratios are used to describe an enzyme's activity toward the metabolism of an administered drug. Plasma drug and metabolite levels, especially collected over time, are more accurate to determine phenotype, but are also more difficult to obtain, in particular in larger population samples. The term *reaction phenotyping* is often used to describe in vitro studies investigating activity toward a drug of interest.	Reviews: PMIDs: 22226243, 17259951, 17273835 Examples: PMIDs: 24218006, 23787463, 22262920, 19519341, 14586384 Review: PMID: 25297949, Examples: PMIDs: 19795925, 15845858
2	Phenotype–genotype	Associations between a phenotype and genetic variation within a single or few genes. Classic phenotype–genotype correlation studies explore the activity between the polymorphisms in a gene (e.g., *CYP2D6*) and the urinary metabolic ratio of the probe drug and its metabolite(s) (e.g., dextromethorphan/dextrorphan).	Examples: PMIDs: 25495411, 23394389, 17971818
3	GWAS	Genome-wide association studies (GWAS) aim to discover an association between common genetic variants and a phenotypic trait, for example, drug response phenotype or adverse drug reaction. See example discussion below.	Reviews: PMIDs: 25582081 [3], 22923055 [5], 20300088 Examples: doi:10.1002/cpt.89, PMIDs: 25350695, 24528284, 19483685, 19706858, 19300499, 18650507

(Continued)

TABLE 11.1 (*Continued*)   Definitions of Terms Central to PGx

#	Method/Strategy	Description	References
4	miRNAs	microRNAs (miRNAs) are small non-coding RNAs that have been shown to be key players in mRNA regulation. This relatively new field is rapidly gaining traction in PGx research. It has been shown that miRNAs contribute to the regulation of many CYP enzymes and drug transporters.	Reviews: PMIDs: 25488579, 24706275, 22510765 Examples: PMIDs: 25802328, 24926315, 24645868, 23935064, 23733276, 22232426, 21457141
5	Epigenetics	Epigenetic alterations (DNA methylation, histone modifications, and chromatin remodeling) represent functionally relevant changes to the genome that do not occur at the nucleotide level. Epigenetic changes have been shown to be associated with expression levels of pharmacogenes.	Reviews: PMIDs: 25677519, 25297728, 24166985, 23935066 Examples: PMIDs: 25071578, 25138234
6	Proteomics	Proteomics is an interdisciplinary field exploring the intracellular protein composition and structure, as well as unique cell proteome signatures. Its application to PGx is just emerging.	Examples: PMIDs: 24830943, 25488931, 25158075, 25218440, 23982336
7	Pharmacometabolomics	Metabolomics at the omics-level is utilized to further our understanding of the mechanisms of drug action and drug response. The identification of metabolic signatures refines phenotype by integrating the impact of PGx, the environment and the microbiome.	http:// pharmacometabolomics. duhs.duke.edu Review: PMID: 24193171 Examples: PMIDs: 25521354, 25029353, 23945822, 23874572

(*Continued*)

TABLE 11.1 (*Continued*) Definitions of Terms Central to PGx

#	Method/Strategy	Description	References
8	PheWAS	Phenome-wide association studies (PheWAS) analyze many phenotypes compared to genetic variants (or other attribute). This method was originally described using electronic medical record (EMR) data from EMR-linked in the Vanderbilt DNA biobank, BioVU, but can also be applied to other richly phenotyped sets. Also see the electronic Medical Records and Genomics (eMERGE) network for additional tools.	http://phewas. mc.vanderbilt.edu PMID: 24270849 Examples: PMIDs: 25074467, 24733291, 24731735, https:// emerge.mc.vanderbilt. edu

metabolism and response (http://pharmacomicrobiomics.com). Saad and coauthors provide a summary on gut pharmacomicrobiomics and review the complex interactions between drugs and microbes (www.gutpathogens. com/content/4/1/16). While the microbiome response-modifying effect has long been appreciated in the fields of nutrition and toxicology, we are only at the beginning to understand the intricate balance between the microbiome and the other omic layers.

Over the past 30 years, PGx has established itself as a research discipline in its own right. It employs many methodological approaches and bioinformatic and biostatistical tools to characterize the contributions of the genome, proteome, transcriptome, and metabolome, that is, the host (patient) phenome, on drug metabolism and response with the ultimate goal to individualize drug therapy.

11.2 METHODS AND STRATEGIES USED IN PGx

This section highlights selected methods and strategies that are utilized in PGx research (not including analyses detailed in other chapters in Section II). Each is accompanied by a brief summary and references.

Although GWAS has many limitations especially for complex relationships [3,5], this method is a mainstay in PGx research and has been successfully applied as demonstrated by the following examples.

11.2.1 Examples of GWAS and Adverse Events

Simvastatin-induced myopathy: Statins are a mainstay therapy to lower low-density cholesterol. In rare cases, myopathy occurs especially when taken in high doses or when taken with certain other medications. This adverse event can lead to rhabdomyolysis, which poses a risk for renal failure and death. Link and coworkers (PMID: 18650507) identified 48 individuals each with definite myopathy and incipient myopathy, respectively, from a large trial of 12,064 participants who had received 20 mg (low) or 80 mg (high) doses of simvastatin. For 85 cases, genomic DNA was available and was subjected to the Illumina Sentrix HumanHap300-Duo BeadChip along 90 controls. Only a single signal with a p-value of less than 10^{-5} was observed. The peak for rs4363657 was found in intron 11 of the solute carrier organic anion transporter gene *SLCO1B1*, the gene encoding the organic anion transporting polypeptide OATP1B1. This efflux transporter is expressed in the liver and facilitates statin uptake (https://www.pharmgkb.org/pathway/PA145011108). This intronic SNP is in near-complete linkage with rs4149056, a nonsynonymous SNP in exon 6 causing a Val^{174}Ala change. The association between myopathy and rs4149056 was replicated in a cohort of individuals (16,643 controls and 21 patients with myopathy) taking a 40 mg dose; the p-value was even smaller in this cohort at 3×10^{-28}. Homozygous carriers of rs4149056 have an almost 20-fold higher risk of myopathy compared to non-carriers. Notably, no associations were found with SNPs in the *CYP3A4* gene, which is a major contributor to simvastatin metabolism.

Floxacillin and drug-induced liver injury (DILI): Floxacillin is a beta-lactam antibiotic agent of the penicillin class known to cause rare, but potentially severe DILI. Daly and coworkers (PMID: 19483685) performed GWAS with the Illumina Human 1M BeadChip and identified a top hit for rs2395029 with a p-value of 8.7×10^{-33} and estimated odds ratio of 45. This SNP, located in the *HCP5* gene, is in complete linkage with *HLA-B*5701* in subjects of European decent. Subsequent direct genotyping of *HLA-B*5701* and a drug-exposed control group affirmed a perfect correlation between rs2395029 and *HLA-B*5701*. Patients with this allele were shown to have an 80-fold increased risk of developing DILI. It needs to be emphasized that this GWAS was carried out on a relatively small number of only 59 patients and 487 controls and a replication cohort of 23 cases. Success of this study was possible due to a single strong signal, but also carefully selected and well-characterized patients. This is a prime example that GWAS study can be performed on relatively small cohorts.

11.2.2 Examples of GWAS and Drug Metabolism (Pharmacokinetic) and Drug Response (Pharmacodynamic) Effects

Clopidogrel metabolism and response: Clopidogrel is commonly prescribed to prevent adverse cardiovascular events after coronary intervention by way of inhibiting platelet aggregation (https://www.pharmgkb.org/pathway/ PA154444041). A subset of patients, however, does not respond to clopidogrel therapy. Using ADP-stimulated platelet aggregation in response to clopidogrel treatment and cardiovascular events as the main outcome measure, Shuldiner and coworkers discovered a highly significant association between non-responsiveness and the *CYP2C* gene cluster (PMID: 19706858). An SNP in linkage disequilibrium with the SNP that defines the non-functional *CYP2C19*2* variant had a *p*-value of 1.5×10^{-13} and accounted for about 12% of the observed variability in ADP platelet aggregation. Furthermore, the authors demonstrated that patients with this variant were at a higher risk of experiencing a cardiovascular event or even death within one year compared to patients without impaired *CYP2C19* function. As described in later literature, clopidogrel requires *CYP2C19*-mediated bioactivation into its active metabolite to exert platelet aggregation (https://www.pharmgkb.org/pathway/PA154424674), corroborating the *CYP2C19* GWAS signal. However, *CYP2C19* variation explains only a fraction of the variability. More recent GWAS work identified an SNP in *PEAR1*, the gene encoding the platelet endothelial aggregation receptor 1 as an additional contributor to treatment response (rs12041331; $p = 7.66 \times 10^{-9}$). This association was observed in patients on aspirin treatment alone or on dual therapy with aspirin and clopidogrel and accounted for approximately 5% of the phenotypic variation. However, Lewis et al. (PMID: 23392654) concluded that it remains to be elucidated whether rs12041331 in *PEAR1* is indeed the causative SNP. Genetic variation in *CYP2C19* and *PEAR1* still explains less than 20% of the variation. As Ritchie et al. discuss [3,5], other associations may exist, but will require more sophisticated analysis tools and/or additional data to capture genotype–phenotype interactions.

Warfarin metabolism and response: Warfarin is one of the most widely used anticoagulants worldwide. It decreases coagulation by inhibiting vitamin K epoxide reductase 1 (VKORC1) and thereby reduces vitamin K-dependent carboxylation of clotting factors. Since warfarin has a narrow therapeutic index, prescribed doses may be too high causing severe and potentially fatal bleeding or too low risking treatment failure. In Europeans, dose requirements vary about 20-fold and about 20% of that variation could be explained by sequence variations in the *CYP2C9* gene, can play an important role in the deactivation of warfarin (https://www.pharmgkb.org/pathway/PA145011113) and its pharmacodynamic effect on VKORC1 (https://www.pharmgkb.org/pathway/PA145011114). To identify additional genetic variation explaining the highly variable warfarin dose requirements, Takeuchi and coworkers (PMID: 2652833) performed the first GWAS study on 1053 Swedish subjects using the Illumina HumanCNV370 BeadChip array. In addition to finding strong associations with *CYP2C9* ($p < 10^{-78}$) and *VKORC1* ($p < 10^{-31}$), only one additional significant association was identified, that is, rs2108622, a nonsynonymous SNP in *CYP4F2* ($p = 8.3 \times 10^{-10}$). While these findings were replicated, this initial GWAS failed to identify additional genetic variants explaining variability in warfarin response. Subsequent efforts in the quest to identify additional genetic as well as non-genetic factors contributing to warfarin response are described in detail in Section 11.4.

11.3 DATABASES AND OTHER RESOURCES

This section highlights selected databases and resources central to PGx research and clinical applications. Relevant PGx databases and resources are described and referenced in Table 11.2.

Note to nomenclature databases: An increasing number of allelic variants were discovered in the 1990s for genes in the CYP superfamily of drug metabolizing enzymes. To keep up with the growing number of variants, a unique nomenclature was first proposed for *CYP2D6* in 1996, which eventually led to the foundation of the Human Cytochrome P450 (*CYP*) Allele Nomenclature Database. Nomenclature for other pharmacogenes and gene families have followed suit, although official nomenclature webpages are only maintained for some (Table 11.2 # 4 and PharmGKB). Briefly, the so-called star (*) nomenclature uses the HUGO-assigned gene symbol, an

TABLE 11.2 Databases and Resources Central to PGx Research and Clinical Applications

#	Database/Resource	Description	Link/References
1	PGRN	The Pharmacogenomics Research Network (PGRN) is a network of scientific groups focused on understanding how a person's genes affect his or her response to medicines through the discovery of novel insights into mechanisms relating genomic variation to differences in drug responses; demonstration of the use and utility of genomic information to improve outcomes for drug therapies and the incorporation of genomic data to predict and personalize medicine use into routine clinical practice.	http://pgrn.org
2	Pharmacogenomics Knowledgebase	The Pharmacogenomics Knowledgebase (PharmGKB) is a comprehensive resource that curates knowledge about the impact of genetic variation on drug response for clinicians and researchers.	www.pharmgkb.org
3	CPIC	The Clinical Pharmacogenetics Implementation Consortium (CPIC) is a shared project between PharmGKB and the Pharmacogenomics Research Network. CPIC's goal is to address some of the barriers to implementation of pharmacogenetic tests into clinical practice. CPIC provides guidelines that enable the translation of genetic laboratory test results into actionable prescribing decisions for specific drugs.	https://www.pharmgkb.org/view/dosing-guidelines.do?source=CPIC PMID: 21270786
4	Nomenclature databases for CYP, UGT, NAT, and TPMT	• The Human Cytochrome P450 (*CYP*) Allele Nomenclature database manages an official and unified allele designation system, as well as the provision of a database of *CYP* alleles and their associated effects. • The UDP-Glucuronosyltransferase (*UGT*) nomenclature page provides *UGT1A* and *UGT2B* haplotype and SNP tables. • The database of arylamine *N*-acetyltransferases (*NATs*) provides annotated information about both human and non-human *NAT* genes and alleles. • The scope of the thiopurine methyltransferase (*TPMT*) nomenclature website is to maintain a common, logical nomenclature system for TPMT alleles in humans. See notes to pharmacogene nomenclature (Section 11.3).	http://www.cypalleles.ki.se PMID: 23475683 http://www.pharmacogenomics.pha.ulaval.ca/cms/ugt_alleles http://nat.mbg.duth.gr http://www.imh.liu.se/tpmtalleles?l=en PMID: 23407052

(*Continued*)

TABLE 11.2 (Continued) Databases and Resources Central to PGx Research and Clinical Applications

#	Database/Resource	Description	Link/References
5	UCSF-FDA TransPortal	Repository of information on transporters important in the drug discovery process as a part of the FDA-led Critical Path Initiative. Information includes transporter expression, localization, substrates, inhibitors, and drug–drug interactions	http://dbts.ucsf.edu/fdatransportal PMID: 23085876
6	ClinVar	The Clinical Variation Resource (ClinVar) is a freely accessible, public archive of reports of the relationships among human variations and phenotypes, with supporting evidence. ClinVar thus facilitates access to and communication about the relationships asserted between human variation and observed health status, and the history of that interpretation.	http://www.ncbi.nlm.nih.gov/clinvar
7	ClinGen	The Clinical Genome Resource (ClinGen) is dedicated to determining which genetic variants are most relevant to patient care by harnessing research data and data from clinical genetics tests being performed, as well as supporting expert curation of these data. ClinGen dosage sensitivity map: The ClinGen consortium is curating genes and regions of the genome to assess whether there is evidence to support that these genes/regions are dosage sensitive and should be targeted on a cytogenomic array.	http://clinicalgenome.org http://www.ncbi.nlm.nih.gov/projects/dbvar/clingen/
8	GTR	The Genetic Testing Registry (GTR) provides a central location for voluntary submission of genetic test information by providers. The scope includes the test's purpose, methodology, validity, evidence of the test's usefulness, and laboratory contacts and credentials. The overarching goal of the GTR is to advance the public health and research into the genetic basis of health and disease.	http://www.ncbi.nlm.nih.gov/gtr
9	LRG	The goal of the Locus Reference Genomic (LRG) database is to generate a system that allows consistent and unambiguous reporting of variants in clinically relevant loci.	http://www.lrg-sequence.org doi:10.1093/nar/gkt1198
10	PACdb	PACdb is a pharmacogenetics-cell line database that serves as a central repository of pharmacology-related phenotypes that integrates genotypic, gene expression, and pharmacological data obtained via lymphoblastoid cell lines.	http://www.pacdb.org PMID: 20216476

(Continued)

TABLE 11.2 (*Continued*) Databases and Resources Central to PGx Research and Clinical Applications

#	Database/Resource	Description	Link/References
11	PheKB	Phenotype Knowledgebase (PheKB) is a collaborative environment to building and validating electronic phenotype algorithms.	https://phekb.org doi: 10.1136/amiajnl-2012-000896
12	DrugBank	The Open Data Drug & Drug Target Database *DrugBank* is a unique bioinformatics and cheminformatics resource that combines detailed drug (i.e., chemical, pharmacological and pharmaceutical) data with comprehensive drug target (i.e., sequence, structure and pathway) information.	http://www.drugbank.ca PMID: 24203711
13	FDA	The U.S. Federal Drug Administration (FDA) maintains a Table of Pharmacogenomic Biomarkers in Drug Labeling.	http://www.fda.gov/drugs/scienceresearch/researchareas/pharmacogenetics/ucm083378.htm
14	I-PWG	The Industry Pharmacogenomics Working Group (I-PWG) is a voluntary and informal association of pharmaceutical companies engaged in research in the science of pharmacogenomics. The Group's discussions, activities, and programs are open and transparent. They are limited exclusively to non-competitive matters.	http://i-pwg.org

Arabic number to indicate the family, a letter that designates the subfamily, and an Arabic number to designate the individual gene within the subfamily. Examples are *CYP2D6* (Cytochrome P450 = superfamily; 2 = family; D = subfamily; 6 = isoform), *UGT1A1*, and *ABCB1B1*. Allelic variants of a gene are distinguished by adding a *number, examples being *CYP2D6*1 and *2*. In most instances, the *1 allele represents the wild-type, or reference, allele; a notable exception is *NAT2* for which *NAT2*4* represents the reference allele. Although this nomenclature is widely accepted in the field and used in the literature, the current nomenclature databases/webpages are being outpaced by the growing numbers of novel variants discovered by next-generation sequencing. There is a desperate need for next-generation-nomenclature systems keeping track of variation in pharmacogenes and their functional consequences to aid clinical implementation of PGx.

11.4 WARFARIN PHARMACOGENOMICS AND ITS IMPLEMENTATION INTO CLINICAL PRACTICE

Warfarin is widely prescribed for the treatment and prevention of thromboembolic disorders and is usually dosed to achieve an international normalized ratio (INR, a measure of anticoagulant activity) between 2 and 3. The risks for bleeding increase significantly with an INR above 4, and the risk for thrombosis increases with INR values below 2. The dose that produces an INR of 2–3 ranges from 0.5 to 10 mg/day or higher. Warfarin is often empirically initiated at a dose of 5 mg/day, with dose adjustment based on INR results. This may lead to supra-therapeutic anticoagulation and bleeding in some patients, while failing to provide adequate anticoagulation in others. While clinical factors such as age, body size, and concomitant medications influence response to warfarin, dose prediction based on clinical factors alone remains poor, especially for patients requiring doses less than 3 mg/day or greater than 7 mg/day (International Warfarin Consortium [IWPC], PMID: 2722908).

Hundreds of candidate gene studies have consistently demonstrated that the *CYP2C9* and *VKORC1* genotypes influence warfarin dose requirements, making warfarin one of the most well-studied drugs in the PGx literature. Genotype also contributes to the risk for major hemorrhage with warfarin, especially in the initial 3 months of therapy (Mega et al., PMID: 25769357). According to data from the IWPC (PMID: 2722908), dosing based on genotype more accurately predicts dose requirements than traditional dosing approaches. There are FDA-cleared platforms for warfarin PGx testing, and guidelines from the CPIC are available to assist with translating genotype results into actionable prescribing decisions (Table 11.2 # 3). For all of these reasons, warfarin is an ideal target for PGx implementation.

Of the variants in CYP2C9, the *2 (Arg144Cys, rs1799853) and *3 (Ile359Leu, rs1057910) alleles are most commonly described and the primary alleles associated with decreased enzyme activity, reduced S-warfarin clearance, and lower warfarin dose requirements in European populations. The CYP2C9*5 (Asp360Glu, rs28371686), *6 (c.817delA, rs9332131), *8 (Arg150His, rs7900194), and *11 (Arg335Trp, rs28371685) alleles occur predominately in populations of African descent and also lead to significant reductions in enzyme activity against S-warfarin and decreased dose requirements. The VKORC1 c.-1639G>A (rs9923231) polymorphism, located in the gene regulatory region, decreases gene expression and further reduces warfarin dose requirements. There are significant differences in CYP2C9 and VKORC1 allele frequencies by race, as shown in Table 11.3.

The FDA-approved warfarin labeling now contains a table with dosing recommendations based on CYP2C9 and VKORC1 genotypes. PGx dosing algorithms that take both genotype and clinical factors into account are also freely available and shown to more accurately predict dose than the table in the warfarin labeling. Two algorithms derived and validated in large populations are described in Table 11.4. A limitation of most dosing algorithms is that they do not include many of the variants that are

TABLE 11.3 Frequencies of Alleles Associated with Warfarin Dose Requirements in Different Ethnic Groups

Allele	Europeans	African Americans	Asians
CYP2C9			
*2	0.13	0.02	ND
*3	0.06	0.01	0.02
*5	ND	0.01	ND
*6	ND	0.01	ND
*8	ND	0.06	ND
*11	ND	0.04	ND
VKORC1-1639A	0.38	0.10	0.91

N/D, not detected.

TABLE 11.4 Warfarin Pharmacogenetic Dosing Algorithms

	IWPC Algorithm	**Gage Algorithm**
Description	Developed by the IWPC, a collaboration of 21 groups from 9 countries and 4 continents who pooled data to create a dosing algorithm with global clinical utility	Derived and validated in a U.S. population. The on-line version includes a dose refinement algorithm that can account for previous warfarin doses and INR response
Derivation cohort	$n = 4043$ (55% White, 30% Asian, 9% Black, 6% other)	$n = 1015$ (83% White, 15% Black, 2% other)
Validation cohort	$n = 1009$ (56% White, 30% Asian, 10% Black, 5% mixed or missing data)	$n = 292$ (83% White, 15% Black, 2% other)
Variables included in the algorithm	Age (in decades), height, weight, VKORC1-1639G>A genotype, CYP2C9 genotype (accounting for *2 and *3 alleles only), Asian race, African-American race, enzyme inducer status (if taking carbamazepine, phenytoin, rifampin, rifampicin), amiodarone use	VKORC1-1639G>A, body surface area, CYP2C9 (*2 and *3 alleles in published algorithm; *2, *3, *5, and *6 included in the on-line algorithm), target INR, amiodarone use, smoking status, African-American race, deep vein thrombosis, or pulmonary embolism
Reference	PMID: 19228618	PMID: 18305455
Website accessibility	http://www.warfarindosing.org	http://www.warfarindosing.org

important in African Americans. As a result, they are significantly less effective at predicting dose requirements in African Americans compared to Europeans (Schelleman et al., PMID: 2538606).

GWAS by Takeuchi et al. (PMID: 2652833) and Cooper et al. (PMID: 2515139) in European populations have confirmed the association between the CYP2C9 and VKORC1 genotypes and warfarin dose requirements. However, with the exception of CYP4F2, they have failed to identify novel associations with warfarin response. In contrast, a GWAS in African Americans, conducted by the IWPC, identified a novel association between an SNP in the CYP2C gene locus and lower dose requirements (Perera et al., PMID: 23755828). The discovery cohort consisted of 533 African Americans from 9 IWPC sites who were genotyped with the Illumina 610 Quad BeadChip or Human1M-Duo 3.0 array. After conditioning on the VKORC1 locus and CYP2C9*2 and *3 alleles, an association emerged with the rs12777823 SNP on chromosome 10 ($p = 1.51 \times 10^{-8}$), which was replicated in an independent cohort of 432 African Americans ($p = 5.04 \times 10^{-5}$). The rs12777823 A allele has a frequency of approximately 20% and is associated with reduced S-warfarin clearance in African Americans.

Another novel variant, located in the folate homeostasis gene folylpolyglutamate synthase (*FPGS*), was identified through whole exome sequencing of samples from 103 African Americans requiring warfarin doses of ≤ 5 or ≥ 7 mg/day (Daneshjou et al., PMID: 25079360). The association between the rs7856096 variant and warfarin dose was replicated in an independent cohort of 372 African Americans, with a dose reduction of nearly 1 mg/day with each minor allele. The risk allele is most prevalent in populations of African descent, with a frequency of 0.15–0.20 in African Americans, but <0.01 in Europeans.

Two randomized, multicenter, controlled clinical trials have assessed the utility of genotype-guided warfarin dosing. The European Pharmacogenetics of Anticoagulant Therapy (EU-PACT) trial compared dosing with a PGx algorithm to a traditional dosing approach (5–10 mg on day 1, then 5 mg per day with dose adjustment based on INR) in 455 patients (Pirmohamed et al., PMID: 24251363). PGx dosing resulted in greater time within the therapeutic INR range, fewer instances of over-anticoagulation, and shorter time to achieve therapeutic dosing. The Clarification of Optimal Anticoagulation through Genetics (COAG) trial randomized 1015 patients to dosing with a PGx algorithm or dosing with an algorithm containing clinical factors only (e.g., age, race, body size, and concomitant medications) (Kimmel et al., PMID: 24251361). In contrast to the EU-PACT trial, the COAG trial found no difference in time in therapeutic range between groups. African Americans, who comprised 27% of study participants, had more INR values above the therapeutic range with PGx dosing.

The EU-PACT and COAG trials only studied the *CYP2C9*2*, **3*, and *VKORC1*-1639G>A variants, which were appropriate for the EU-PACT trial, in which 98% of participants were European. However, recent data suggest that failure of the COAG trial to account for African-specific variants, such as *CYP2C9*5*, **6*, **8*, **11*, and the rs12777823 SNP, may have contributed to the higher risk for supratherapeutic dosing with the PGx algorithm in the African American cohort (Drozda et al., PMID: 25461246). A third trial is ongoing and is examining the effect of genotype-guided dosing on risk for venous thromboembolism and major bleeding in patients taking warfarin after major orthopedic surgery.

Based on substantial and consistent data that genotype influences warfarin dose requirements, genotype-guided dosing became the standard of care for patients newly starting warfarin at the University of Illinois Hospital in 2012. This serves as one of the earliest examples of clinical implementation of warfarin PGx [7]; the process for providing genotype-guided dosing is outlined in Figure 11.1. The initial dose, prior to the return of genotype results, is calculated with a clinical dosing algorithm embedded in the electronic health record. Genotype results are targeted to be available prior to

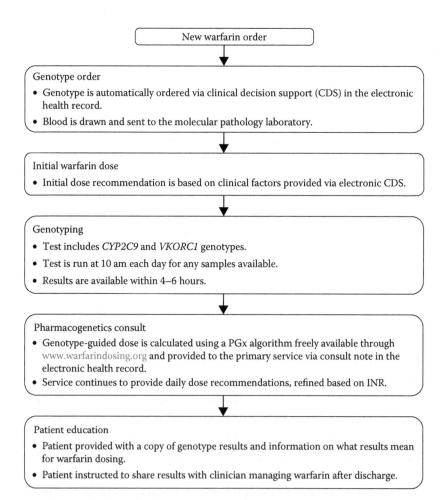

FIGURE 11.1 Process for providing genotype-guided dosing at the University of Illinois Hospital & Health Sciences System.

the second warfarin dose, at which time a PGx service provides a genotype-guided dose recommendation. The service continues to provide daily dose recommendations, refined based on INR response to previous doses, until the patient reaches a therapeutic INR or is discharged. Outcome data with the service are expected, which in addition to data from the ongoing clinical trial, should help guide the future of warfarin PGx.

11.5 PHARMACOGENOMICS KNOWLEDGEBASE— PharmGKB

Overview: The Pharmacogenomics Knowledgebase (PharmGKB, Table 11.2 # 2) is a web-based, publically available resource (funded by the NIH and NIGMS). PharmGKB's mission is to support the understanding of how genetic variation contributes to differences in drug metabolism and response. The knowledge contained in PharmGKB is human-curated from a variety of sources to capture the relevant pharmacogenomic relationships among genes, drugs, and diseases. Controlled vocabularies are imported from trusted repositories such as the Human Genome Nomenclature database (gene symbols and names; http://www.genenames.org/), Drugbank (drug names and structures; Table 11.2 # 12), MeSH (http://www.nlm.nih.gov/mesh/MBrowser.html), and SnoMed (http://www.ihtsdo.org/snomed-ct/) (disease terminology). The PharmGKB resource is free for all, but subject to the usage agreement for research purposes only without redistribution.

Curated knowledge: The PharmGKB Knowledge Pyramid (Figure 11.2) illustrates the available knowledge and how these diverse types of information are integrated in several PharmGKB features. At the foundation of the

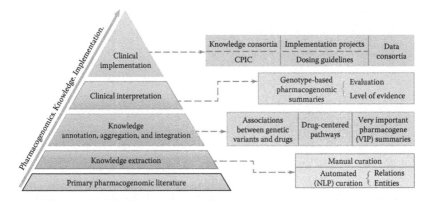

FIGURE 11.2 PharmGKB knowledge pyramid. CPIC, Clinical Pharmacogenetics Implementation Consortium; NLP, natural language processing.

knowledgebase is the pharmacogenetic and pharmacogenomic literature, which is annotated, aggregated, and integrated in gene variant annotations, drug-centered pathways, and summaries of important pharmacogenes.

The core components of PharmGKB are variant annotations, which extract the association between a single variant (polymorphisms or haplotype) and a drug phenotype reported in a published article. Accompanying study parameters such as study size, population ethnicity, and statistics are recorded for each association. Clinical annotations combine multiple variant annotations for a variant–drug–phenotype association to create a summary report per applicable variant genotype. As shown in Figure 11.3, the

FIGURE 11.3   Clinical annotation for rs4244285 and clopidogrel. Screenshot of the summaries capturing the association between rs4244285 and clopidogrel by rs4244285 genotype based on annotated evidence.

underlying evidence that supports the clinical annotation is displayed as a list of single variant annotations with links to the individual articles. The associations are reported in a relative fashion as compared to other genotypes.

Each clinical annotation is assigned a level of evidence score. The score is a measure of the confidence in the association by a PharmGKB curator. It is based on several criteria including replication of the association and *p*-value. The criteria underlying the scoring of the evidence level are previewed and published [8] and summarized at https://www.pharmgkb.org/page/clinAnnLevels.

Knowledge about a specific drug or gene might be also condensed in drug-centered pathways or in very important pharmacogene (VIP) summaries. A pathway consists of a diagram and description highlighting genes involved in the pharmacokinetics or pharmacodynamics of a drug or drug class. The relationships illustrated are based on published literature and referenced in the description and components tab of the pathway page. The pathways are manually constructed as graphic images, converted to gpml (GenMapp pathway markup language) and BioPAX formats (http://www.biopax.org/) to be stored in the underlying database. A VIP summary provides a pharmacogenetic-focused overview based on the literature about an important gene involved in drug responses. The summary covers background information about variants and haplotypes, if applicable. Currently, 54 VIP summaries (https://www.pharmgkb.org/search/browseVip.action?browseKey=vipGenes) and 110 pathways (https://www.pharmgkb.org/search/browse/pathways.action) are available on PharmGKB. Many of the pathways and VIP summaries are written in collaboration with external experts in the field and published in the journal *Pharmacogenetics and Genomics.*

Clinical implementation of pharmacogenomic knowledge is part of PharmGKB's efforts (Figure 11.2). PharmGKB supports several clinically related projects including data-sharing and implementation projects (https://www.pharmgkb.org/page/projects). The CPIC (Table 11.2 # 3) publishes gene–drug dosing guidelines to enable physicians to incorporate knowledge about a patient's genetics in the drug selection/dosing decision in cases where the patient's genotype is available (Relling et al., PMID: 21270786 and Caudle et al., PMID: 24479687). The guidelines are drafted in a standard format according to standard operating procedures, published in *Clinical Pharmacology and Therapeutics* and simultaneously posted to PharmGKB (additional details are provided in Section 11.6).

Knowledge Organization: PharmGKB serves a broad user base with pharmacogenomics relationships for research interests and knowledge about clinically applicable gene–drug relationships contributing to the vision of personalized medicine. The general drug page layout is shown in Figure 11.4. The knowledge is organized under tabs for clinical pharmacogenomics, pharmacogenomics research, overview, properties, pathways, related genes, drug and diseases, and downloads/link-outs. Gene and disease pages are similarly organized.

The clinical pharmacogenomics tab is divided into dosing guidelines, drug labels, and clinical annotations. The dosing guidelines tab includes the posting of genotype-based drug dosing guidelines including those by CPIC and the Royal Dutch Association for the Advancement of Pharmacy Pharmacogenetics Working group (Figure 11.4a). The extended guideline feature allows the user to access available CPIC guideline recommendations specific to a genotype selected in a dropdown menu (Figure 11.4b).

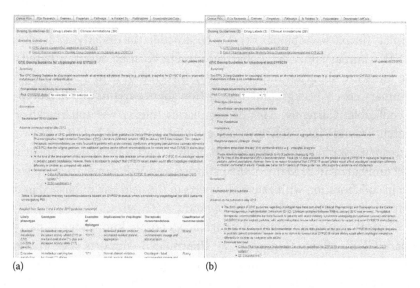

(a) (b)

FIGURE 11.4 PharmGKB clopidogrel drug page. The clopidogrel drug page showing tabs for clinical pharmacogenomics (dosing guidelines, drug labels, and clinical annotations), pharmacogenomics research, overview, properties, pathways, related genes, drug and diseases, and downloads/link-outs. (a) Part of the excerpt from the CPIC guidelines for *CYP2C19* genotypes and clopidogrel therapy (Scott et al., PMID: 23698643). (b) Recommendations for *CYP2C19*2/*2* genotype based on Table 2 from the CPIC guidelines for *CYP2C19* genotypes and clopidogrel therapy. CPIC, Clinical Pharmacogenetics Implementation Consortium; *CYP2C19*, cytochrome P450, family 2, subfamily C, polypeptide 19.

The drug labels tab provides excerpts from the drug label and a downloadable highlighted label PDF file for drug labels containing pharmacogenomics relevant information approved by the U.S. Food and Drug Administration (FDA), European Medicines Agency, or the Pharmaceuticals and Medical Devices Agency, Japan. Clinical annotations (described under *curated knowledge*) for the drug are listed under the clinical pharmacogenomics tab sorted by variants. The pharmacogenomics research tab contains genomic variants or gene haplotypes related to drugs with links to the variant annotations (described under *curated knowledge*) that capture the association from individual articles. The overview and properties page have basic information about the drug. The pathway tab links to available curated pharmacokinetic or pharmacodynamic pathways related to the drug. Related genes, drug, and diseases are compiled from different knowledge pieces in PharmGKB, which are indicated through different symbols, supporting a relationship of the drug with a gene and/or disease. Download/link outs connect to external resources and vocabulary.

11.6 CLINICAL PHARMACOGENETICS IMPLEMENTATION CONSORTIUM

CPIC is part of the mission of the Pharmacogenomics Research Network (PGRN) (Table 11.2 # 1) and supported and housed by PharmGKB. The incorporation of genomic data into routine clinical practice is the ultimate goal of PGx, and a number of challenges and barriers have, however, been identified to hamper efforts. Among the top identified challenges were difficulties interpreting genotype tests and how to translate genotype information into clinical action. To address these challenges, CPIC is devising guidelines that are designed to help clinicians understand how available genetic test results may be used to tailor drug therapy for the individual patient. The guidelines follow standardized formats (Caudle et al., PMID: 24479687), are peer-reviewed, are freely available, and are updated approximately every two years. Candidate gene/drug pairs for guideline development are selected based on the levels of evidence supporting recommendations in favor of choosing a different drug or altering dose. Each guideline provides a wealth of information including extensive literature reviews and summaries of the supporting literature, information of genetic variation, and dosage recommendations based on genotype-predicted phenotype. To address a growing interest in informatics aspects of CPIC guidelines and clinical implementation of PGx, the CPIC Informatics Working Group is tasked to identify and resolve technical barriers to the implementation of the guidelines within a clinical electronic environment and to create comprehensive translation tables from genotype to phenotype to clinical recommendation. Workflow diagrams and example clinical decision support alerts and consults are now included into the guidelines as resources facilitating the incorporation of pharmacogenetics into an electronic health record with clinical decision support. Information contained within the guidelines are also increasingly disseminated with other organizations and databases including the GTR, PGRN, and eMERGE listed in Table 11.1 # 8, but also www.guidelines.gov, the NHGRI Genomic Medicine Working Group, the Institute of Medicine Genomic Medicine Roundtable, the American Medical Informatics Association, and the FDA.

To date, guidelines have been published for 15 gene/drug pairs. These include the *SLCO1B1*/simvastatin, *CYP2D19*/clopidogrel, and *CYP2C9* and *VKORC1*/warfarin, which have been highlighted as GWAS examples.

REFERENCES

1. Bertino, J. S. Jr., DeVane, C. L., Fuhr, U., Kashuba, A. D., Ma J. D. (eds.). (2012) *Pharmacogenomics: An Introduction and Clinical Perspective*, McGraw-Hill, New York.
2. Johnson, J. A., Ellingrod, V. L., Kroetz, D. L., Kuo, G. M (eds.). (2015) *Pharmacogenomics Application to Patient Care*, 3rd edition, American College of Clinical Pharmacy, Lenexa, KS.
3. Ritchie, M. D., Holzinger, E. R., Li, R., Pendergrass, S. A., and Kim, D. (2015) Methods of integrating data to uncover genotype-phenotype interactions. *Nature Reviews Genetics* **16**, 85–97.
4. Daly, A. K. (2010) Genome-wide association studies in pharmacogenomics. *Nature Reviews Genetics* **11**, 241–246.
5. Ritchie, M. D. (2012) The success of pharmacogenomics in moving genetic association studies from bench to bedside: Study design and implementation of precision medicine in the post-GWAS era. *Human Genetics* **131**, 1615–1626.
6. Wang, L., McLeod, H. L., and Weinshilboum, R. M. (2011) Genomics and drug response. *The New England Journal of Medicine* **364**, 1144–1153.
7. Nutescu, E. A., Drozda, K., Bress, A. P., Galanter, W. L., Stevenson, J., Stamos, T. D., Desai, A. A. et al. (2013) Feasibility of implementing a comprehensive warfarin pharmacogenetics service. *Pharmacotherapy* **33**, 1156–1164.
8. Whirl-Carrillo, M., McDonagh, E. M., Hebert, J. M., Gong, L., Sangkuhl, K., Thorn, C. F., Altman, R. B., and Klein, T. E. (2012) Pharmacogenomics knowledge for personalized medicine. *Clinical Pharmacology and Therapeutics* **92**, 414–417.

## CHAPTER 12

Exploring De-Identified Electronic Health Record Data with i2b2

Mark Hoffman

CONTENTS

12.1 INTRODUCTION

The Precision Medicine Initiative announced by President Obama in his 2015 State of the Union Address envisions a large cohort of patients whose biological and clinical data are utilized to improve the personalization of health care [1]. The electronic health record (EHR) is widely recognized as the richest source of clinical information about patient phenotype and will be an important source of medical data for the Precision Medicine Initiative. The "Meaningful Use" section of the American Reinvestment and Recovery Act (ARRA) of 2009 provided funding to support the widespread adoption of EHRs. Meaningful Use also provided a phased set of capabilities that are required of EHR systems in order to qualify for federal funding. These requirements have promoted the generation of data that is optimized for analysis by rewarding the use of discrete data capture fields and standardized terminologies such as the Systematized Nomenclature of Medicine—Clinical Terms (SNOMED-CT) and Logical Observation Identifiers Names and Codes (LOINC).

Discrete codified data representing most aspects of clinical workflow can be used to stratify patients at a highly granular level, a necessary capability for precision medicine. Research performed using this granular information can provide useful insights into complex diagnostic and prognostic associations. As the adoption of EHRs is extended to include genetic information [2], research to examine associations between DNA variants and a wide variety of phenotypes will become feasible. Examples of phenotypes that can be investigated include individual risk of developing disease, personal variability in drug response, or markers that are prognostic of cancer progression. Patient outcomes can often be inferred from EHR-derived data using both direct and indirect measures. Direct outcomes can include discharge status, for example, mortality, or readmission with a diagnosis that indicates the failure of a previous treatment. Indirect measures of patient outcomes can include lab results for biomarkers such as hemoglobin A1c or high-density lipoprotein levels.

Before discussing applications that enable research with de-identified data generated from EHRs, it is important to review the reasons why such research cannot easily be supported directly through the EHR. While EHRs are extremely useful for improving the quality of health care, they were designed and implemented to support the delivery of patient care, not research. As such, key features that would have made research using direct access to EHR data feasible are not widely included in EHRs. Privacy requirements, including those codified in the Health Information Portability and Accountability Act (HIPAA), limit access to protected health information (PHI). While EHRs provide all of the capabilities necessary to comply with privacy regulations during the delivery of patient care, they generally lack features that enable compliance with privacy regulations for research purposes. All EHRs provide the ability to generate complex reports that are utilized to support clinical operations. While many of the fields associated with PHI can be hidden from these reports using filters, obscuring date of service is problematic and requires additional functionality. Many categories of research, including health services research, require understanding of the time between events. EHRs typically do not include an embedded visual query builder that does not require advanced training to be productive. In the absence of a query builder, the remaining approach is for the small group of experts in information technology (IT) group to receive requests from researchers specifying the data that they need to be extracted from the EHR.

The limited availability of IT personnel trained to write queries against data in the EHR is a significant barrier to this approach to research. This group focuses on the success of the EHR for patient care and managing the daily operations of the EHR. Research queries often have to be worked into their queue to perform as their schedule permits. It is not uncommon for the request from a researcher to require clarification and multiple iterations before a useful file can be generated. Furthermore, performing direct queries against the EHR database can introduce performance risks to the patient care setting. For example, a query that interrogates a table used for medication orders could slow the responsiveness of computerized provider order entry (CPOE). The Informatics for Integrating Biology and the Bedside (i2b2) initiative was launched to provide researchers with a user-friendly application to enable queries against de-identified information derived from the EHR [3].

Many significant national initiatives have recognized the importance of linking EHR data with genomic data. For example, the Electronic Medical Records and Genomics (eMerge) project includes organizations such as Northwestern University, Vanderbilt, the Mayo Clinic, Group Health, Boston Children's Hospital, the Marshfield Clinic, and Geisinger Health system. These organizations use EHR systems from multiple vendors to address a variety of significant clinical research questions. The eMerge initiative has yielded new insights into obesity, juvenile arthritis, and the genetic basis of response to chemotherapy drugs. A key challenge with eMerge has been extracting information from EHRs. Often participating sites work through data warehouses instead of direct connections to the EHR systems to accomplish their goals.

At the University of Missouri—Kansas City (UMKC), a recently formed group, the Center for Health Insights (CHI) focuses on implementing and developing informatics resources to promote biomedical research. The CHI team includes experts in data science, genomics, medical informatics, statistics, *Big Data*, and bioinformatics. The CHI works closely with UMKC clinical affiliates, including Children's Mercy Hospital and Truman Medical Center, to assist their efforts to make effective and efficient use of i2b2, REDCap, and other research informatics platforms. The structure of the CHI, in which clinicians, biomedical researchers, and engineers work closely together, provides a model for successful research informatics support.

This chapter summarizes representative use-case scenarios in which i2b2 can accelerate research. The key elements of the i2b2 informatics architecture will be reviewed and then a brief tutorial describing the workflow of a typical analysis will be provided.

12.2 APPLICATION

Biomedical researchers working in healthcare settings frequently comment about frustration at limited access to aggregate clinical data. The barriers are often related to resource limitations, especially the staffing of the IT team. While organizations differ in their interpretations of HIPAA, policies implemented to protect patient privacy are also viewed as a barrier to accessing clinical data for research purposes. By offering an accessible user-friendly application that accesses de-identified data, i2b2 offers a resolution to both of these barriers.

Since its inception, i2b2 has been applied to many areas of biomedical research. Formal projects aligned with the i2b2 organization are identified as *disease-based driving biology projects* or DBPs. Clinical investigators associated with i2b2 DBPs gain prioritized access to informatics experts in the i2b2 organization.

There are numerous applications of i2b2, but this chapter will highlight four applications:

1. Feasibility analysis for cohort discovery

2. Formation of research networks to scale beyond individual institutions

3. Longitudinal analysis using data extracted from i2b2

4. Integration with bio-repositories to support biological analysis

12.2.1 Feasibility Analysis

Biomedical researchers face numerous challenges in performing human subject's research. Failure to recruit an adequate number of subjects for a research study is identified as one of the most common causes of failure for a clinical research study. A 2011 report from the Tufts Center for the Study of Drug Development comments that as many as two-thirds of sites in a study fail to meet their recruitment targets. Often clinical researchers overestimate the number of patients in their clinic or practice who would meet the inclusion criteria for a clinical protocol, introducing delays into the completion of the protocol.

One of the most valuable applications of i2b2 is to enable feasibility analysis or cohort screening. Using i2b2, a researcher can easily query de-identified data to determine how many patients possess the attributes in the inclusion criteria for their study. Likewise, the logical filters of i2b2 can support exclusion logic. For example, a researcher planning a study related to drug metabolism may want to exclude patients with a documented diagnosis code for cirrhosis.

12.2.2 Formation of Research Networks

Many research studies require more participants than are likely to be found within a single institution. The Patient Centered Outcomes Research Institute (PCORI) has funded clinical research network development, including some that are based on i2b2 [4]. The ability to integrate recruitment efforts across sites is another benefit of i2b2, though complex data governance, legal, and financial issues continue to make these initiatives challenging.

Shrine is the official data-sharing network framework for the i2b2 community [5]. It provides the technical architecture to connect multiple i2b2 instances and is the basis for a number of emerging research networks. Under Shrine, data remain local but federated querying capabilities are implemented to support networked research.

12.2.3 Longitudinal Analysis

EHRs include a variety of information that can inform and improve research. One of the most significant opportunities in EHR-based research is the opportunity to monitor patient interactions with the healthcare system over time in a longitudinal analysis. While health care claims data sets generated by payor data can provide longitudinal data, EHR-derived data offer many advantages for longitudinal data.

As with payor data, most EHR systems include patient diagnosis code, documented using International Classification of Disease-9 (ICD-9) codes. All EHR vendors are preparing for ICD-10 which will offer even more granularity. Within i2b2, ICD codes are often used as one of the first filters in defining a query. Unlike claims data, EHR systems also include information related to laboratory orders and results, medication orders, date and time stamped procedure events, vital sign documentation, and often billing data. The availability of these additional categories of data over multiple patient encounters is one of the key strengths of EHR-based data analysis. I2b2 allows users to define queries that apply inclusion or exclusion logic based on any of the categories of data and then can generate a data set for further analysis.

12.2.4 Integration with Bio-Repositories

For genomic research, there is a significant need to query clinical data, find patients who meet the inclusion criteria, and then access a biological sample (blood or tissue) from those individuals. Those samples can then serve as the source of DNA, RNA, or protein analytes for further inquiry. Some implementations of i2b2 support this critical workflow by capturing a flag in their EHR to document that a patient has provided consent to be included in a biobank [6] (Table 12.1).

These i2b2 capabilities will become recognized as an important part of fulfilling President Obama's 2015 State of the Union announcement that Precision Medicine is a national priority.

It is important to acknowledge the limitations of i2b2. The database schema for i2b2 is optimized to favor a user-friendly interface and limits the ability to perform complex queries directly within i2b2. Formal data warehouses using relational database structures or emerging *Big Data* architectures such as Hadoop will continue to play an important role in biomedical research. Also, despite the improvements in the use of discrete fields to store information in the EHR, text documents continue

TABLE 12.1   Representative i2b2 Projects

Project	Organization	Approach
Rheumatoid arthritis	Brigham and Women's, Partners Healthcare	Use i2b2 to identify genes and DNA variants associated with susceptibility to rheumatoid arthritis
Great Plains Research Collaborative	University of Kansas (lead)	Research network
Airways disease	Harvard	Longitudinal analysis
Consented biorepository	Cincinnati Children's Hospital	Linkage of EHR-derived records to consented biobank specimens

to store a significant portion of clinical information in the EHR. While some i2b2 research has been performed utilizing natural language processing (NLP) of textual clinical information [7], there are not yet widely reliable methods to fully de-identify text documents which can include PHI such as surnames, telephone numbers, or addresses [8]. Therefore, research utilizing the extraction of text documents into i2b2 will continue to require institutional review board (IRB) oversight.

12.3 DATA ANALYSIS

Moving data from an EHR into i2b2 requires a general workflow common to all EHR vendors and platforms and similar to the extract, transform and load (ETL) process common to data warehouse implementations. The sequence of these steps may vary slightly across implementations, as can the locations in which the data transformations occur (locally or in a cloud hosted environment). Decisions about who performs an i2b2 implementation can have significant impact on the features and flexibility of the system in use by any particular organization as well as the support available for troubleshooting.

12.3.1 Extract Data from EHR Database

EHR systems utilize database systems to store a wide variety of information, including orders, results, diagnosis codes, and other information generated during the delivery of patient care. While Meaningful Use has promoted standards that enable interoperability, each EHR vendor maintains proprietary backend database systems. The first step in moving data into i2b2 is to extract the data from the EHR database. This requires in-depth knowledge of the table or file structure and system architecture of the EHR platform.

12.3.2 Account for Privacy Considerations

Most i2b2 implementations utilize de-identified information, requiring that the data extracted from the EHR be processed to remove or obfuscate any of the 18 data elements protected by HIPAA. Date fields must be removed or shifted using an algorithm that cannot be reversed. In order to retain value for research, it is important that dates be shifted consistently so that the time between events for a person represented in i2b2 can be accurately evaluated even if the exact dates of service are unknown. For example, it is highly significant whether a patient is readmitted for a complication 1 week or 3 months after a surgery. Likewise, date shifting approaches that can preserve day of the week effects are preferable to random shifting as utilization of many healthcare resources changes during weekends. Some i2b2 implementations can transfer identified information from the EHR into i2b2, and these require additional data handling processes and appropriate IRB oversight.

12.3.3 Data Mapping

I2b2 uses a platform-specific ontology to standardize terms. The i2b2 ontology uses parent–child relationships to enable hierarchical queries. For example, the high-level category of "Microbiology Cultures" would have children that include "Blood Cultures" and "Urine Cultures." A query using the "Microbiology Cultures" term would retrieve all cultures including blood and urine, while a query in which the user specifies blood cultures would only retrieve that specific group of orders. The raw data extracted from the EHR must be mapped to this ontology for i2b2 to function. When standardized terminologies such as LOINC, SNOMED-CT, and ICD-9 are utilized, this is a relatively clean process. Data that are not associated with one of these terminologies must be mapped through an additional batch or curation process.

12.3.4 Data Transformation

Every EHR vendor has a proprietary data schema. I2b2 uses a data schema that is optimized to support simple interactive queries. A data transformation process converts the mapped data from the raw EHR format into the i2b2 structure. Software scripts that can be run automatically facilitate these transformations. Some institutions pass their data through a data warehouse before transforming it into i2b2.

12.3.5 Data Load

Once the data have been de-identified, mapped, and transformed, it is loaded into the i2b2 installation. Some organizations use local hardware computing platform managed on site to store the data. Other institutions use a private, secure, cloud hosted model.

While these steps are common to all i2b2 installations, there are a variety of strategies to implement i2b2. Selection of the approach to i2b2 deployment is often influenced by budget and staffing. The degree to which an organization wants independence in managing and modifying the system is another key factor. The level of support expected by users and tolerance for system down time are further elements. Three i2b2 implementation strategies are prevalent.

1. *Independent implementation*: The first approach is an independent process that is managed by the research organization and their clinical collaborator. They are responsible for each of the five processes identified above. This approach can require significant time and effort. Support is expected exclusively from the internal staff. Organizations taking this approach have complete control over the configuration of their i2b2 system.

2. *Consultant driven*: The second model is to engage consultants to perform the majority of the process, though local stakeholders will have decisions and responsibilities at many key milestones. The consultants assist with the technical work and provide guidance on data governance and other policies. They also contribute to the data mapping and standardization. These implementations often include the option to subscribe to expert support. In this model, the consultants may not have access to the technical infrastructure hosting i2b2, requiring that the research organization provide technical troubleshooting support. The consultant-driven approach offers the advantage of utilizing best practices but can often have high costs. The Recombinant division of Deloitte consulting is one of the most widely utilized i2b2 consulting services.

3. *EHR vendor facilitated*: The third model, utilized by two of the UMKC healthcare partners, Children's Mercy Hospital and Truman Medical Center, is to apply the services of their EHR vendor Cerner® Corporation in implementing i2b2. This model also requires involvement of the clinical team at the research organization but utilizes standardized extraction, transformation, and loading processes. In this model, the system can be hosted in a private cloud, limiting the hardware investment required. The vendor can provide both technical and application support. A limitation of this model is that the researchers are generally not able to independently install plug-ins or make modifications to the backend database.

As an example, two of the healthcare provider organizations that partner with UMKC have utilized the Cerner implementation of i2b2, which falls under the third model. Cerner provided professional services to support the five steps of i2b2 implementation. They supply cloud-based hosting from their data center and offer continuous support. Data are loaded from the EHR into i2b2 on a regular schedule using automated processes. The research organizations were actively involved in the data mapping work and in validating that the data in i2b2 were an accurate representation of the information found in the EHR. Validation includes activities such as confirming that the number of cases of patients with a particular diagnosis code in i2b2 corresponds with the number of cases found with a query run direction against the EHR. Another organization in the Kansas City Area, the University of Kansas Medical Center, utilizes Epic® as their EHR. They pursued a successful i2b2 implementation process that was largely independent, with limited use of consulting support [9].

Once the data are loaded into i2b2, it can then be accessed for research. Most interaction with i2b2 occurs through the user-friendly front-end application. This provides the ability to easily select inclusion or exclusion criteria using either a *drill-down* approach through the i2b2 ontology or a search capability. Multiple formats are available for the output of a query, including summary reports and a timeline view. The ability to export data from i2b2 in delimited format provides the ability to perform statistical analysis using standard applications such as SAS or R.

While i2b2 is optimized for healthcare data, it has informed initiatives that serve as templates for the integration of genomic information with the EHR [10]. As an open source application framework, i2b2 has fostered a development ecosystem that has generated plug-ins and optional features. Figure 12.1 shows a representation of i2b2 Core Cells, optional cells, workbench plug-ins, the web client, and a CRC plug-in to support patient counts.

The ability to incorporate optional features or develop new capabilities is a strong benefit of i2b2.

12.4 TUTORIAL

Using i2b2 is relatively simple compared to many analytical applications. Users have a front-end program that uses drag and drop inclusion or exclusion criteria and customizable filters. In this tutorial, we will follow a researcher who wants to determine whether or not their institution has at least 50 patients who will meet the inclusion criteria for a project under consideration. Dr. Jones is an expert in pediatric epilepsy and wants to

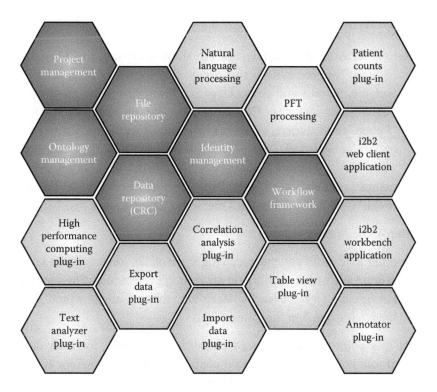

FIGURE 12.1   i2b2 features (from i2b2.org).

know whether there is a genetic basis for outcomes variability in response to tegretol. She wants to focus on children between the ages 6 and 12.

Step 1: *Launch i2b2*

Dr. Jones launches i2b2 from a web browser. She selects the appropriate environment, enters her username and password, and views the opening screen of the application.

Step 2: *Select Demographic Criteria*

In the i2b2 home screen, Dr. Jones choose the "Navigate Terms" tab. She then selects the "Demographics" folder and expands it. After picking the "Age" folder, two groups that are relevant to her research are displayed—"0–9 years old" and "10–17 years old." She expands both of these folders. From the 0–9 folder, she chooses 6, 7, 8, and 9 and then drags them to the Query Tool pane and repeats that process for the 10, 11, and 12 year olds from the 10–17 group.

Step 3: *Select Diagnosis Code(s)*

Returning to the "Navigate Terms" tab, she now expands the "Diagnoses (Billing)" option. Within this group, she finds the "Diseases of the Nervous System and Sense Organs." From there, she chooses the "Other Disorders of the Central Nervous System" category and finds "Epilepsy and recurrent seizures." She can either use the entire group or narrow her search criteria to only those patients with "Generalized convulsive epilepsy," "Epilepsy unspecified," or another related category. She chooses "Generalized convulsive epilepsy" and drags it to the "Group 2" pane of the Query Tool (Figure 12.2).

Step 4: *Medication Filter*

In order to identify patients receiving tegretol, Dr. Jones uses the search prompt instead of drilling down through the menus. She sees four options for tegretol (carbamazepine), chooses all four, and drags them to Group 3 in the Query Tool.

At this point, Dr. Jones could establish other filters for each query group. For example, she could confine the date parameters. Instead she chooses to use all available data.

Step 5: *Run Query*

Dr. Jones clicks the "Run Query" button at the bottom of the Query Tool. She can choose the type of query results from this list of options:

- Patient list

- Event list

- Number of patients

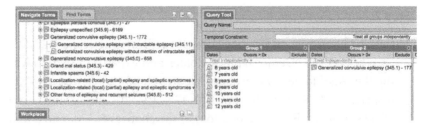

FIGURE 12.2   i2b2 query tool.

- Gender patient breakdown

- Vital status patient breakdown

- Race patient breakdown

- Age patient breakdown

- Timeline

She chooses the patient list, event list, number of patients, gender patient breakdown, and timeline options. Then, she launches her query. After a brief wait, her results are returned.

Step 6: *Review Results*

After running her query, the results are displayed in a summary format. Only four patients fulfilled her criteria, leading Dr. Jones to realize that she needs to modify her criteria, add additional sites to her study, or identify an alternative. She could consider using Shrine to collaborate with other sites and develop a research network.

This example demonstrates how a researcher can use i2b2 and within minutes perform a rapid feasibility analysis. There are factors that users must take into consideration when making decisions based on i2b2 results. For example, there can be data that are relevant to feasibility that is not stored in the EHR. In this example, if the medication list is only generated from inpatient prescribing, access to outpatient tegretol prescriptions might have changed the results.

Querying de-identified clinical data is a powerful technique to inform biomedical research. Many of the barriers to performing research with clinical data are overcome by i2b2. This application enables researchers who are not trained in query languages or command line applications to easily access EHR data with a user-friendly front end. Through simple *drag-and-drop* interactions and easily configurable filters, users can design and run a query in minutes. The de-identification of EHR data used in i2b2 mitigates barriers related to concern about PHI. By resolving two key barriers to performing research with EHR-derived biomedical data, i2b2 opens up new avenues for inquiry to a much larger community. As an open source application, i2b2 has a vigorous and active development community generating new plug-ins and enhancements. I2b2 will play an important role in enabling access to clinical and phenotypic data as the Precision Medicine Initiative moves forward.

REFERENCES

1. Collins FS, Varmus H. A new initiative on precision medicine. *The New England Journal of Medicine.* 2015;372(9):793–5.
2. Hoffman MA. The genome-enabled electronic medical record. *Journal of Biomedical Informatics.* 2007;40(1):44–6.
3. Murphy SN, Mendis ME, Berkowitz DA, Kohane I, Chueh HC. Integration of clinical and genetic data in the i2b2 architecture. *AMIA Annual Symposium Proceedings.* 2006;2006:1040.
4. Waitman LR, Aaronson LS, Nadkarni PM, Connolly DW, Campbell JR. The greater plains collaborative: A PCORnet clinical research data network. *Journal of the American Medical Informatics Association.* 2014;21(4):637–41.
5. Weber GM, Murphy SN, McMurry AJ, MacFadden D, Nigrin DJ, Churchill S et al. The Shared Health Research Information Network (SHRINE): A prototype federated query tool for clinical data repositories. *Journal of the American Medical Informatics Association.* 2009;16(5):624–30.
6. Marsolo K, Corsmo J, Barnes MG, Pollick C, Chalfin J, Nix J et al. Challenges in creating an opt-in biobank with a registrar-based consent process and a commercial EHR. *Journal of the American Medical Informatics Association.* 2012;19(6):1115–18.
7. Yang H, Spasic I, Keane JA, Nenadic G. A text mining approach to the prediction of disease status from clinical discharge summaries. *Journal of the American Medical Informatics Association.* 2009;16(4):596–600.
8. Murphy SN, Gainer V, Mendis M, Churchill S, Kohane I. Strategies for maintaining patient privacy in i2b2. *Journal of the American Medical Informatics Association.* 2011;18(Suppl 1):i103–8.
9. Waitman LR, Warren JJ, Manos EL, Connolly DW. Expressing observations from electronic medical record flowsheets in an i2b2 based clinical data repository to support research and quality improvement. *AMIA Annual Symposium Proceedings.* 2011;2011:1454–63.
10. Masys DR, Jarvik GP, Abernethy NF, Anderson NR, Papanicolaou GJ, Paltoo DN et al. Technical desiderata for the integration of genomic data into electronic health records. *Journal of Biomedical Informatics.* 2012;45(3):419–22.

## CHAPTER 13

Big Data and Drug Discovery

Gerald J. Wyckoff and D. Andrew Skaff

CONTENTS

13.1 INTRODUCTION

The last 30 years have led to a sea change in how drug discovery is performed. Advances in both chemistry and biology have led to rational drug design, improvements in medicinal chemistry, and new methods for both target selection and lead compound generation. However, in spite of all these advances, only five of 40,000 compounds tested in animals reach human testing and, shockingly perhaps, only one in five that reach clinical trials is approved [1]. Moving these compounds through the early stages of development represents an enormous investment of resources: both financial and human resources, as well as the opportunity costs of the research and time investment. While the advancements within both chemistry and biology generally demand the most attention, it is changes in computation that perhaps will yield the most fruit, and the *Big Data* challenge for drug discovery is to move the target and lead identification and validation phase of drug discovery into a fast and accurate in silico mode, from its previous slow and costly chemical screening mode. With the National Institutes of Health (NIH) launching discovery portals for drug discovery [2], Google collaborating with Stanford to deliver new networking technology [3], and Oxford recently launching their own initiative [4] backed by both government and private sources, all centered around *Big Data*, it's clear that in silico drug discovery is moving into a new phase. In this chapter, we'll discuss what tools exist for in silico work, how you can use them, and how *big* the data can get.

Advances in both combinatorial chemistry and laboratory automation have allowed high-throughput screening of compounds to become common methodologies, and this has led to the wide availability of libraries of molecules for drug discovery [5–7]. A better understanding of the chemistry behind small-molecule docking and examples of ligands which have binding activity to different targets, in parallel with the explosion of available tertiary and quaternary protein structures, has enabled in silico modeling of small molecules to become a standard practice in both academic and commercial laboratories. However, this method has commercial successes that are primarily limited to *me too drugs* of successfully marketed pharmaceuticals. The statin drug family, which lower cholesterol by targeting the enzyme HMG-CoA reductase, is a primary example of this—there are more than 10 marketed statins, all of which share structural similarity in the HMG moiety. With this mode of thinking, designing a successful new drug relies on identifying compounds directed against known targets, not on new human disease gene targets. Docking methods attempt to identify the optimal binding position, orientation, and molecular interactions between a ligand and a target macromolecule. Virtual screens, using large local computers or cloud-based resources, can screen compound structure libraries with millions of entries and yield hit rate of several orders of magnitude greater than that of empirical, bench-based screening for a fraction of the cost. Focused target and small molecule libraries can be generated, focusing on either specific properties of small molecules, molecules that contain a specific substructure (e.g., a binding pocket or active site), or combinations of both. Structural fingerprinting can extend this concept by encoding a three-dimensional structure in a bitwise representation of the presence or absence of particular substructures within a molecule. Algorithms can also be used for clustering of structures within a chemical library, and other metrics allow determination of small-molecule subsets that are filtered by desired properties for drug-like, lead-like, or fragment-like compound representations for in silico screening.

The primary purpose of small-molecule-based drug discovery is finding new chemical entities targeted against human disease-related proteins or targeted to block pathogen activity. Current efforts are greatly assisted by a variety of new techniques; large human genome variant databases, genome-wide association studies (GWAS), and finer dissection of disease haplotypes and haplotype mapping efforts. Finding a disease target, however, does not assure a viable drug target is in hand; in fact, the descriptors that define drug targets are often antithetical to the descriptors that define disease genes. In particular, for example, targets of successful drugs are often much more conservative than disease genes from an evolutionary perspective—likely due to the need for animal trials leading to the Food and Drug Administration (FDA) clinical trial process.

Increasing pharmaceutical budgets have not assured that new chemical entities will succeed; in fact, FDA approval has been on a steady decline. Many new drugs were introduced in the 1990s to treat previously untreated or undertreated conditions, but the pace of introduction has declined since 2000—in most years, back to levels not seen since the 1980s. The introduction of priority drugs—those that, according to the FDA, provide a *significant therapeutic or public health advance*—has also slowed, from an average of more than 13 a year in the 1990s to about 10 a year in the 2000s. In the mid-2000s, product patent expiration threatened about $40 billion in U.S. pharmaceutical revenue, or about $80 billion worldwide. If smaller, "non-blockbuster" drugs replace this, we'd need between 80 to 100 drugs in the U.S. and about double that worldwide, if average revenue was $500 million per drug. In tandem with the (primarily pharmaceutical driven) push to discover NMEs, there has been a (primarily research driven) push to discover new disease targets [1,8].

13.2 IN SILICO SCREENING FOR DRUG DISCOVERY

With these pressures in place, the need for faster, cheaper, and higher-quality methods for drug discovery and development become apparent. Development pipelines, which can take up to 15 years, do not allow for the rapidity of development that is necessary. As the approval process is relatively fixed once a molecule is at the stage of being an investigational new drug, an obvious place to attack with Big Data methods is the hit-to-lead phase of drug discovery.

This is where in silico methods become our most viable means of screening. When paired with in vitro high-throughput screening methods, this greatly shortens screening times for hits, improves the hit-to-lead cycle in drug development, and enhances the ability of drugs to successfully be brought to market. The drug discovery process is typically at least a decade long governed in part by the FDA's need for multiple animal trials prior to clinical trials in humans. Safety and efficacy are, of course, the watchwords of this process. How, then, does in silico screening enhance this process?

13.3 OUTLINE OF IN SILICO RESOURCES AND METHODOLOGIES

When discussing in silico methods, it's necessary to consider the scope of methods and resources available for the researcher. This section deals with what is available, as of the publication of this book, for helping to both find targets and screen for small molecule compounds in the context of the average academic research lab.

13.3.1 Target Utilities

Consider that without protein targets, most drug screening cannot proceed. Therefore, the most likely start to a drug discovery project is with a target that is identified as part of a screen already being carried out within a lab—indeed, ideally, one would be able to predict efficacy and side effects from such screens [9]. Typically, a researcher is working on one gene or pathway—or, alternatively, on a disease and comes across a protein that might be considered to be a target for drug development. Selection of drug targets is inherently biased, then, by what is already being examined. Interestingly, the genes that are identified as being causal for, or associated with, a disease are often much more rapidly evolving, in terms of the nonsynonymous change rate, than those genes coding for proteins of successful drug targets [10]. This might have to do with the need to develop successful animal models with which to test a drug, or perhaps it is due to alternative factors such as the likelihood of having a good protein structure model. Regardless, the choice of target greatly affects the success of drug development efforts and is often overlooked—in theory, any protein could be a target—so choosing a target more carefully might be seen as a waste of time. However, knowing those resources that might help you vet a target more thoroughly is worthwhile.

13.3.1.1 RCSB

Perhaps the most venerable website is that at www.rcsb.org, the Protein Data Bank. This repository of protein structure data was established at Brookhaven National Laboratory in 1971. Updated weekly with new structural information, the PDB file format contains exhaustive information not only about the structure of a protein, as experimentally determined by a researcher, but also about the composition of the protein, how the structure was generated, resolution, and information on the models. It is the most likely source for an initial protein structure with which a docking/modeling experiment will be performed. With over 35,000 distinct protein structures—deposited by researchers worldwide—familiarity and use of this resource is practically mandatory for drug development researchers.

13.3.1.2 Variant Libraries

Another key issue is variants that may be clinically relevant in protein targets. For this, three different sources are recommended. The first is another respected and long-standing site: OMIM, or Online Mendelian Inheritance in Man, at http://www.omim.org/. OMIM started as a book compiled by the late Victor McKusick, devoted to cataloging of Mendelian traits and variants in diseases. It has become the lead repository of information that relates genotype to phenotype for human genes, with information on over 12,000 genes. When examining a potential drug target, it's absolutely worth discovering if a clinically relevant variant exists and is annotated within OMIM before proceeding.

The second resource is dbSNP, located at http://www.ncbi.nlm.nih.gov/SNP/. dbSNP records information on single-nucleotide polymorphisms (SNPs), small insertions and deletions (indels), small copy number variants (CNVs), short tandem repeats (STRs), and microsatellites. As intended, these variants are not only in the coding region of proteins, and thus may be related to expression. Considering which variants exist in and around the gene that codes for a protein of interest may give insight into how broadly applicable a drug designed for that target may be.

Finally, the Gene Expression Omnibus (GEO) at http://www.ncbi.nlm.nih.gov/geo/ contains gene expression variants catalogued from a large number of expression studies, particularly eGWAS data. Knowing where and when a drug target is expressed is likely key for successful development, dosing, and deployment of potential drugs.

13.3.1.3 HapMap

The International HapMap project, http://hapmap.ncbi.nlm.nih.gov/, is another resource to consider. This project is a compilation of chromosomal regions containing shared genetic variations across humans, the resource is often used for the discovery of disease genes—some of which may be potential drug targets. Similar to dbSNP, it's worth considering what variants are present in a gene region which codes for a target protein before proceeding with drug development.

13.3.2 Small Molecule Utilities

Once a target is identified and a structure has been obtained (either through modeling or experimentally), in silico drug discovery would proceed by having a set of small molecules for screening. The resources below are good starting places for gathering compound libraries and, in some cases, for ordering compounds for in vitro or in vivo testing of lead compounds.

13.3.2.1 National Cancer Institute

A good place to start is at the National Cancer Institute of the National Institutes of Health. NCI maintains stocks of compounds that can be used for early stage drug screening. These sets are chosen to represent diverse compounds (the so-called diversity sets) or for mechanistic screens of function (mechanistic sets) or, more broadly, sets of approved drugs are maintained. In particular, the diversity sets (which as of publication contain about 1600 molecules) are a good place to start screening. These sets of diverse compounds with relatively interesting Lipinski "Rule of five" characteristics are also available for physical screening. The diversity set is curated from a library of a much larger library (140,000 compounds) and representative compounds from structural subsets of the larger screen are selected based on the quantity available for researchers looking to follow up on successful hits. In silico hits within these sets can be rapidly followed up at the bench for this reason, and hit expansion can proceed from these analyses. For a primer on what NCI has available, go to http://dtp.nci.nih.gov/index.html. A similar and larger set of data is also available from the Molecular Libraries Program at http://mli.nih.gov/mli/compound-repository/mlsmr-compounds/.

13.3.2.2 ZINC

Probably the largest set of information on compounds available is located at http://zinc.docking.org. ZINC, which stands for "ZINC is not commercial" [11], archives structures and characteristics of well over 20 million small molecule compounds. While many of these are similar to one another, it is a truly invaluable resource for folks in the drug discovery field. Importantly, subsets of structures of molecules for screening with certain characteristics can be created and downloaded easily. Restrictions include size, charge, number of hydrogens, and compound availability, to name just a few. Caveats include that, as of this writing, searches might vary as compounds become available (or unavailable) from different vendors, which can lead to frustration if you're unaware of this. Downloading the information you need for a screen and creating a control version for internal use in your lab over the course of a drug screen might be worthwhile.

13.3.2.3 Commercial Vendors

A variety of commercial vendors have compounds available for screening and, in many cases, make basic compound composition/structure information available. While there are too many to name, several may be useful and these include Enamine (http://www.enamine.net/), Chembridge (http://www.chembridge.com/index.php), and ChemDiv (https://chemdiv.emolecules.com/). Vendors may need to manufacture molecules, and shipping times differ; plan accordingly if you have a tight time frame for your screening.

13.3.3 Biologics and Utilities

13.3.3.1 Biologics

Not all drugs, of course, are small molecules. In silico algorithms for the development of biologics are becoming available and a number of these allow engineering of biologics in silico. Currently, much of the software available is process management software meant to monitor or integrate information from analytics readouts from, for example, Biacore or other assays.

13.3.3.2 Utilities

If drug screening and development is a primary goal, utilities and resources for maintaining databases of chemical libraries, drawing Markush structures, and other every-day tasks are likely a necessity. While there is too much to discuss here, ChemAxon (https://www.chemaxon.com/) is one vendor that provides a unified platform for the sorts of activities that become incredibly tedious to do using, for example, spreadsheets in a lab environment. Notably, the JChem Suite is a good start if you are performing large numbers of in silico analyses—suitable for a *Big Data* deployment. A version of JChem, JChem for Office, allows integration of small molecules structures and property calculations into Word and Excel documents—a highly useful product both for managing experiments and for writing up publication-quality papers and patentable structures. Worth noting is that there are a large variety of scripts available online that can assist researchers in automating the process of small molecule screening—we discuss several in this article but an excellent source of these is often the user groups related to particular docking software.

13.3.4 Docking/Modeling

There are a wealth of docking and modeling programs that are available for use. We later extensively detail how to use AutoDock Vina [12] (http://vina.scripps.edu/) and PyMol [13] (http://www.pymol.org/) for modeling and visualization, but the combination of these with AutoDock Tools in a Linux environment is a likely combination of screening tools in an academic setting, and we'll cover this scenario later. Of course, other highly useful and respected programs include CHARMM (Chemistry at Harvard Macromolecular Mechanics [14]) (http://www.charmm.org/), AMBER (Assisted Model Building with Energy Refinement) (http://ambermd.org/), and AutoDock 4 (http://autodock.scripps.edu/), all of which implement force field models [15,16] for performing docking and modeling software and are distributed under various licenses that generally allow free academic use (with citation in publications). Following the same mode, Glide from Schroedinger is a commercially available and supported program. A new implementation from Schroedinger, QM-Polarized Ligand Docking, uses quantum mechanical calculations, specifically ones that help deal with polarization in molecules, that when married to Glide can theoretically achieve higher accuracy, especially for problematic molecules—and it is far from the only program to perform these functions. Many of these programs require a local server to use effectively, which means that you must be able to build and maintain the server as well as install and update all necessary software. While this is certainly not difficult, it is not necessarily within the purview of labs looking to do relatively small numbers of docking/modeling experiments. Publicly accessible servers that implement these methodologies offer an alternative: a good example is SwissDock (http://www.swissdock.ch/), which implements the EADock DSS software, an algorithm which uses CHARMM energies for docking. A commercial server, http://www.dockingserver.com/web (from Virtua Drug Ltd), offers both free and pay services for performing docking on their servers, and there are other such services in the market.

While setting up a local server might seem unnecessary for small numbers of experiments, being able to rerun and rescore results (using, for example, NNScore 2.0 or other software) may make the option seem more palatable [17–19]. While the *barrier-to-entry* for such work may seem high, our tutorial in this chapter should enable you to get started relatively rapidly. From a Big Data perspective, a local implementation might not suffice for storage or computational reasons. An alternative might be implementation of these algorithms within a cloud-computing environment, such as sold by Amazon Web Services (AWS), Google, or Microsoft. Such implementations have numerous advantages—low initial overhead, high potential for scaling, redundancy, and high availability to name a few. Settling on an implementation and feature set that works for your research is key [20,21]. There are several recent comparisons of features of docking and modeling software, and they would serve as good reviews before jumping into this work.

13.3.5 Fingerprinting/Structural Comparison

Another function that is likely necessary is structural comparison of small molecules, or fingerprinting of ligands and, potentially, protein surfaces. Fingerprinting abstracts the *shape* of small molecules, generally into numeric strings that are faster to compare than three-dimensional shapes or molecular force fields [22]. Fingerprinting is implemented in some cases to facilitate comparison of ligands, proteins, or interaction surfaces using Tanimoto coefficients, which allow for distance matrices of interactions to be built and examined. PLIF, or *protein–ligand interaction fingerprinting*, is implemented in PyPlif (https://code.google.com/p/pyplif/) and examples can be seen in that work. Fingerprints of proteins, implemented, for example, as PubChem substructure fingerprints (https://pubchem.ncbi.nlm.nih.gov/), allow comparison of proteins or substructures—some implementations, such as ChemVassa [23], allow direct comparisons of proteins or substructures with ligands or drug-like small molecules. Structural comparison can be used as a component of hit expansion, or for determining what library of small molecules to use for docking experiments.

13.3.6 Visualization

Visualization of results is an important step. Probably the most highly used program is PyMol, mentioned above, which is free for non-commercial use. However, other programs exist and have strong followings. JMol is one such program (http://jmol.sourceforge.net/), another is MarvinSpace (http://www.chemaxon.com/products/marvin/marvinspace/). RasMol (http://rasmol.org/) is also worth noting for its feature set and relative ease of use. Cost and features vary widely. It's worth noting that visualization is widely absent in the *Big Data* space. Visualization of large sets of comparative docking results across a set of different targets, for example, is largely absent—as are tools that specialize in visualizing unwanted or *off-target* effects.

13.3.7 Others

Not all functionality can be neatly fit into a particular bin. OpenBabel is one tool that is nearly required to utilize any of the methods mentioned, as it allows translation between different structural file formats. However, it does more than this and is often the basis for other scripts and tools that allow rapid deployment of docking technologies. NNScore (1.0 or 2.0) [18,19] is a rescoring algorithm that enables either custom-built neural networks for rescoring or a pre-built (but accurate) network designed to overcome some of the inherent problems in molecular modeling software. There are a variety of other rescoring utilities—it's worth noting that it is also possible to tweak the scoring methods inside many programs (such as AutoDock Vina) to weight particular CHARMM forces differentially from standard behavior in the program.

13.4 TUTORIAL UTILIZING THE ANDROGEN RECEPTOR AS AN EXAMPLE

In order to gain a deeper appreciation for exactly how the above methodologies apply, we'll work through an example of an in silico docking screen for the androgen receptor.

Step 1: The first step is to have a tertiary structure representation for the target molecule you wish to screen. We'll work with the androgen receptor from *Homo sapiens* for this example. The best source for experimentally determined three-dimensional protein structures is, of course, RCSB Protein Data Bank (PDB) [24], located on the web at http://www.rcsb.org/pdb/home/home.do (all links and protein/ligand identities are current as of the date of publication). Using the search function for androgen receptor pulls out a list, and this can be sorted by species as well as other qualities. Reading through the available structures, we can quickly find 4OEA, which is the Human Androgen receptor binding domain in complex with dihydrotestosterone. It's worth noting at this point that targets with a bound ligand, especially their native substrate, work best for screening—as will be seen later, this helps determine the binding box that will be screened. We can then download the file in PDB format and prepare the receptor.

Step 2: While it might be tempting to first prepare your receptor, it's worthwhile now to consider the source of the ligands you'll use for screening. If you have a library of small molecules that you wish to screen at a later point, you might, for example, want to perform an in silico screening at this point to determine what hits you expect to see. Alternatively, you might have no compounds that you've worked with yet, and therefore, you're using the in silico screen to guide your later bench screening. For the purposes of this example, we'll assume that this example in silico screening is your first experimental protocol, and we'll use it to derive a set of small molecules that you need to order later for bench testing. If that is the case, consider the source of your in silico library. An excellent resource is ZINC (discussed above and re-linked here). Located at http://zinc.docking.org/, this website houses information on the chemical properties and structures of millions of small molecule compounds. The "subsets" tab at the top of the page allows you to find a set of predetermined subsets of ZINC that can be downloaded and used for in silico screening. Importantly, all of the subsets have been determined to contain small molecules that are *purchasable*, which means there is some commercial source for the compound. A scan of the table shows what criteria were used for filtering. For the purposes of this example, the "In Stock," "Lead-like" compounds, called "Leads now" is likely a good start: it has over 3.5M small molecule compounds, filtered to have a narrow molecular weight range. Clicking the link allows you to download the files for your platform, and we'll utilize MOL2 format for now as it makes later steps simpler. Clicking the link provides a script that performs the download (actually, several downloads in serial), so make sure to save and run the script. The MOL2 files downloaded from ZINC are 600 mb each with close to 150,000 compounds in the single file. Typically, a Perl script is used to break them into individual files before running the ligprep script. An example of this script is

```
perl separe_ligands.pl -f 21_p0.1.mol2
```

where separe_ligands.pl was a file containing a rather lengthy script that can be found at www.biostars.org/p/9011/ as the 4th entry on that webpage. This script actually assigns the ZINC identifier as the name of the individual MOL2 file as it is being created.

The file 21_p0.01.mol2 was one of the 600 mb files from ZINC.

Step 3: At this point, you should prepare your receptor for screening. For purposes of this example, we'll utilize AutoDock Vina on an Ubuntu Linux platform for this work. However, it is sometimes easier to prepare and visualize the receptor on your local workstation, which may be Windows based. Regardless, the steps are similar (though not identical). Vina is available here: http://vina.scripps.edu/download.html, and it is recommended to download a pre-compiled executable unless you are an expert user or are trying to perform specialized functions. You'll also need MGLTools which can be downloaded here: http://mgltools.scripps.edu/downloads, and, once again, a pre-compiled binary is recommended. For setting up the receptor, you need to download and set up both Vina and MGLTools and then launch MGLTools. At this point, you read in the molecule (4OEA.pdb) and ensure that only the receptor (NOT the ligand) is selected. Remove any other chains besides the one that you want to dock against. At this point, select the chain, delete water molecules, and add hydrogens from the EDIT drop down menu (gray bar). Then, go to GRID in the submenu window (tan) and Macromolecule, Choose... and select your PDB file. You will be able to save as a PDBQT file, the proper format for further docking. MOL2 files contain hydrogens and this procedure also sets the Gasteiger charges necessary for docking. We'll discuss setting the grid box later.

Step 4: Preparing your ligands. You need to automate this as much as possible, particularly for a large number of potential ligands. It's worthwhile to consider if you want to build a subset of your larger library for screening, particularly on a local computer. Screening a very large number of files without some parallel deployment is not recommended. The available script to perform this, prepare_ligand4.py, is a python script that requires python to be downloaded onto your local machine. Downloading this script shows a number of options and explains why MOL2 files are preferred. Ideally, this script is further automated with a shell script in Linux that reads all potential ligand files, performs the necessary ligand preparation, and writes them. Make sure to add hydrogens and compute charges; a sample script and options are provided here (see box, Batchprep script). It is run as:

```
./batchprep.sh *.mol2
```

Batchprep script
```
for file in *;
do ~/MGLTools-1.5.6/M0047LToolsPckgs/
AutoDockTools/Utilities24/prepare_ligand.py -l
$file;
done
```

Where batchprep is a file that contains the script in the box at left. Note the location of your script may be different depending upon your installation. This may take some time to run, depending on your processor. The script could theoretically be threaded across a number of processors, split, for example, by compound name, and this would speed up conversion considerably especially on a large computer or deployed cloud cluster. Ideally, your ligands will be named according to their ZINC numbers and will thus be ZINC0911876. pdbqt or similar. Note the name and prefix, as it will be useful later.

Step 5: Looking at your receptor, you need to determine two additional sets of parameters. In short, you need to look at where your /initial/ ligand was bound and determine if that location seems reasonable for further docking efforts. If so, you draw the target box enclosing that ligand area (usually with some additional space around it). You can then determine the coordinates of the center of your binding box and the size of that box. The center (X, Y, Z) coordinates correspond the receptor you've already prepared, and the size is in angstroms. Note the center and size, and you need to make a configuration file for this. In our case, the file, saved as conf.txt, appears in the box (titled Configuration File).

Configuration File:
```
receptor = 4OEA.pdbqt

center_x = 2
center_y = 6
center_z = -7

size_x = 15
size_y = 15
size_z = 15
num_modes = 9
```

Then you need to create a script to perform your screening. This assumes that all of your ligands are prepared and in the SAME DIRECTORY as your receptor and configuration file—see the box titled Screening Script.

Screening Script:
```
#! /bin/bash
for f in ZINC_*.pdbqt; do
    b=`basename $f .pdbqt`
    echo Processing ligand $b
    vina --config conf.txt --ligand $f --out ${b}
out.pdbqt --log ${b}log.txt
done
```

This script will process all of your ZINC ligands against the configuration file you set up for your receptor and will put all of the poses and logs into the same directory. Ideally, you could then head the log file and grep the file for all poses; this would be sorted into a script which could be used to find all particular small molecule ligands that bind with lower than a specified energy (say, −9.0 or lower). Particularly for a large screen, this will likely result in a large number of compounds that match.

Step 6: Once you have a list of potential small molecule binders, you'll recover the files (perhaps, the 10 best molecules from your screen) and look at the poses that performed best. In this case, we'll look only at the top two poses. One of these is shown to be a molecule that is known to bind androgen receptor; Tetrahydrogestrinone THG, or "the clear," notable for being used in doping in sports and having been linked to previous investigations. At this point, the ideal method would utilize PyMol (available here: http://www.pymol.org/) to visualize the dock, and hydrogen bonds would be examined as well as the fit of the small molecule into the docking box. ZINC would be queried to determine which vendors sold compounds for screening; in this case, only some of the compounds would be available, and these could be ordered for bench screening.

Further steps from here likely are determined by your experimental interest. If a large number of screens are being performed, a rescoring method that tends to match your ligands to likely binders more closely could be used. NNScore 2.0, which was discussed previously, is a good rescoring tool—but there are others. Another direction would be to utilize a fingerprinting tool to profile the top 10 hits and utilize these to find an enriched set of potential binders (e.g., from within ZINC) and perform another in silico screen as described above on this set of potential binders. Some fingerprinting methods can potentially screen millions of compounds in a day, even on a desktop computer, and would thus be good companions for a Big Data approach.

At the end of this example, you can see several themes emerge: the need, ideally, for a large computational deployment—potentially involving cloud-based resources; the need for new methods of rapid screening that generate novel ligand sets for in silico screening; and the need for researchers to have facility with scripting and command-line procedures to enable rapid screening within the Big Data world of drug discovery.

REFERENCES

1. DiMasi, J.A. et al., Trends in risks associated with new drug development: Success rates for investigational drugs. *Clin Pharmacol Ther*, 2010. **87**(3): pp. 272–7.
2. Vaughn, P., NIH-led effort launches Big Data portal for Alzheimer's drug discovery. 2015, NIH: http://www.nih.gov/news/health/mar2015/nia-04.htm.
3. Ramsundar, B. et al., Massively multitask networks for drug discovery, in *Proceedings of the International Conference on Machine Learning*, Lille, France, 2015.
4. Griffin, F., Oxford launches $140M big data drug discovery initiative, in *HealthTech Zone*. 2013, TMCnet. p. 1.
5. Ma, X., Z. Wang, and X. Zhao, Anticancer drug discovery in the future: An evolutionary perspective. *Drug Discov Today*, 2009. **14**: pp. 1136–1142.
6. Stumpfe, D., P. Ripphausen, and J. Bajorath, Virtual compound screening in drug discovery. *Future Med Chem*, 2012. **4**(5): pp. 593–602.
7. Shoichet, B., Virtual screening of chemical libraries. *Nature*, 2004. **432**(7019): pp. 862–5.
8. DiMasi, J.A., The economics of new drug development: Costs, risks and returns, in *New England Drug Metabolism Group Spring Meeting*. 2008, NEDMDG, Cambridge.
9. Keiser, M.J. et al., Predicting new molecular targets for known drugs. *Nature*, 2009. **462**(7270): pp. 175–81.
10. Zhang, Q. et al., Selective constraint: A hallmark of genes successfully targeted for pharmaceutical development. *Am J Drug Disc Dev*, 2012. **2**: pp. 184–93.
11. Irwin, J.J. and B.K. Shoichet, ZINC—A free database of commercially available compounds for virtual screening. *J Chem Inf Model*, 2005. **45**(1): pp. 177–82.
12. Trott, O. and A.J. Olson, AutoDock Vina: Improving the speed and accuracy of docking with a new scoring function, efficient optimization, and multithreading. *J Comput Chem*, 2010. **31**(2): pp. 455–61.
13. PyMOL, PyMOL: A user-sponsored molecular visualization system on an open-source foundation. 2010, Schroedinger Corporation. http://www.schrodinger.com/products/14/25/.
14. Brooks, B.R. et al., CHARMM: A program for macromolecular energy, minimization, and dynamics calculations. *J Comput Chem*, 1983. **4**: pp. 187–217.

15. Roterman, I.K. et al., A comparison of the CHARMM, AMBER and ECEPP potentials for peptides. II. Phi-psi maps for N-acetyl alanine N'-methyl amide: Comparisons, contrasts and simple experimental tests. *J Biomol Struct Dyn*, 1989. 7(3): pp. 421–53.

16. Roterman, I.K., K.D. Gibson, and H.A. Scheraga, A comparison of the CHARMM, AMBER and ECEPP potentials for peptides. I. Conformational predictions for the tandemly repeated peptide (Asn-Ala-Asn-Pro)9. *J Biomol Struct Dyn*, 1989. 7(3): pp. 391–419.

17. Durrant, J.D. et al., Comparing neural-network scoring functions and the state of the art: Applications to common library screening. *J Chem Inf Model*, 2013. S3: pp. 1726–35.

18. Durrant, J.D. and J.A. McCammon, NNScore 2.0: A neural-network receptor-ligand scoring function. *J Chem Inf Model*, 2011. 51(11): pp. 2897–903.

19. Durrant, J.D. and J.A. McCammon, NNScore: A neural-network-based scoring function for the characterization of protein-ligand complexes. *J Chem Inf Model*, 2010. 50(10): pp. 1865–71.

20. Rehr, J. et al., Scientific computing in the cloud. *Computing in Science and Engineering*, Jaunary 13, 2010. Jan.

21. Hazelhurst, S., Scientific computing using virtual high-performance computing: A case study using the Amazon Elastic Computing Cloud. *Proceedings of the South African Institute of Computer Scientists and InformationTechnologists Conference*, Stellenbosch, South Africa, 2008.

22. Tovar, A., H. Eckert, and J. Bajorath, Comparison of 2D fingerprint methods for multiple-template similarity searching on compound activity classes of increasing structural diversity. *ChemMedChem*, 2007. 2(2): pp. 208–17.

23. Moldover, B. et al., ChemVassa: A new method for identifying small molecule hits in drug discovery. *Open Med Chem J*, 2012. 6: pp. 29–34.

24. Berman, H.M. et al., The protein data bank. *Acta Crystallogr D Biol Crystallogr*, 2002. 58(Pt 6 No 1): pp. 899–907.

## CHAPTER 14

Literature-Based Knowledge Discovery

Hongfang Liu and Majid Rastegar-Mojarad

CONTENTS

14.1 INTRODUCTION

In the past decade, advances in high-throughput biotechnology have shifted biomedical research from individual genes and proteins to entire biological systems. To make sense of the large-scale data sets being generated, researchers must increasingly be able to connect with research fields outside of their core competence. In addition, researchers must interpret massive amounts of existing knowledge while keeping up with the latest developments. One way researchers cope with the rapid growth of scientific knowledge is to specialize which leads to a fragmentation of scientific literature. This specialization or fragmentation of literature is a growing problem in science, particularly in biomedicine. Researchers tend to correspond more within their fragments than with the field's broader community, promoting poor communications among specialties. This is evidenced within the citations of such literature as authors tend to heavily cite those within their narrow specialties. Interesting and useful connections may go unnoticed for decades. This situation has created both the need and opportunity for developing sophisticated computer-supported methodologies to complement classical information processing techniques such as information retrieval.

One methodology to the above problem is literature-based discovery (LBD) which directly addresses the problems of knowledge overspecialization. LBD strives to find connections that are novel and that have not been previously explicitly published. In 1986, Don Swanson presented his first literature-based hypothesis that fish oil may have beneficial effects in patients with Raynaud's syndrome. Literature search identified 2000 articles on Raynaud's syndrome and around 1000 articles on dietary fish oil. The two groups of articles appear to be isolated but have significant common attributes related to blood viscosity, platelet aggregability, and vascular reactivity. The possibility of linking fragmented literature through intermediate or shared attributes has commonly been described as Swanson's ABC model (Figure 14.1). The model can be implemented as two discovery processes. In the first, open discovery, concepts A, B, and C are known and the relationship between A and B and B and C are known,

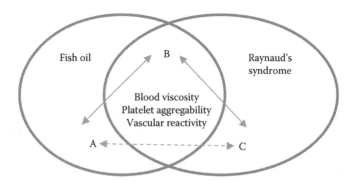

FIGURE 14.1 Literature-based discovery (LBD) methodology overview. The solid links between the concepts are known facts (relations) from the literature. The dashed link is a hypothetical new relation (discovery).

but the relationship between A and C has not been identified. The goal is to discover that relationship. Discovery is facilitated by using particular B concepts to draw attention to a connection between A and C that has not been previously noticed. Starting with concept A, all the B concepts related to A are retrieved and then all C concepts related to B are also retrieved. If there are no prior reported relations between A and C, then a hypothesis of association between A and C can be formulated which can be confirmed or rejected through human judgment, laboratory methods, or clinical investigations. The second paradigm, closed discovery, is used to explain an observed phenomenon. A relationship between, two entities provided by user, A (e.g., flecainide) and C (e.g., heart failure), is assumed but undefined. Proving the existence of the relationship is facilitated by identifying B (e.g., kcnip2 gene) concepts that provide mechanistic links between A and C.

14.2 TASKS IN LBD

LBD deploys general text mining techniques such as named entity recognition or information extraction and attempts to combine mined facts into serendipitous and novel hypotheses. Because many combinations of mined facts are possible, LBD applies various techniques to filter and prioritize possible hypotheses for further investigation. Common tasks in LBD are described in the following sections.

14.2.1 Term Recognition

The first necessary step toward LBD is to recognize terms representing concepts or entities in free text. Approaches for the term recognition task can be categorized into three main types: (1) rule/pattern-based recognition methods characterized by handcrafted name/context patterns and associated rules, (2) dictionary lookup methods requiring a list of terms, and (3) machine learning methods utilizing annotated corpora. Among them, machine learning methods have achieved promising performance given a large annotated corpus. The availability of machine learning software packages has boosted the baseline performance of the task. In the biomedical domain, multiple annotated corpora have been developed and made publicly available over the years for detecting genes/proteins, diseases, or chemical entities. The state-of-the-art performance for term recognition has been comparable to human experts with accuracy in the range from 80% to over 90%.

14.2.2 Term Normalization

To recognize terms representing the same concept or entity and to facilitate the integration of knowledge captured in various knowledge bases or databases for knowledge discovery, it is also desired to map terms in text to ontologies or entries in databases. The task is usually referred as term normalization. The approach for term normalization can be divided into several steps: (1) establishing a mapping table which maps terms in text to ontology concepts or database entries, (2) handling lexical variations and synonyms, and (3) resolving ambiguities when one term can be mapped to multiple concepts or entries. The state-of-the-art performance for term normalization has also been comparable to human experts with accuracy in the range from 60% to 80%.

14.2.3 Information Extraction

After term recognition and/or normalization in literature, simple co-occurrence-based approaches can be applied by assuming the co-occurrence relationship indicates certain associations. To narrow the discovery to certain associations among certain concepts or entities, event or relation extraction has been incorporated into LBD. Many such information extraction systems have been developed ranging from event extraction such as detecting genetic mutations or protein posttranslational modifications to relation extraction such as gene–gene interactions or gene–disease associations. For example, the Chilibot system extracts relationships between genes, chemicals, and diseases and visualizes these relationships in a network of nodes with edges indicating the type and direction of the relationship. It is possible to look for nodes that are not directly connected but have one (or more) intermediate node(s) that are connected to the disconnected ones.

14.2.4 Association Mining and Ranking

To keep meaningful co-occurrence associations, different association mining and ranking techniques have been applied (Yetisgen-Yildiz and Pratt 2009). Popular association mining metrics include the following:

- *Association rules*: These rules were originally developed with the purpose of market basket analysis where a market basket is a collection of items purchased by a customer in a single transaction. An association rule can be interpreted as two items tend to be purchased together in a single transaction. Let Dx be the collection of documents containing x.

The two important measures for associations between two concepts, X and Y, are as follows: support ($|D_X \cap D_Y|$, the number of documents containing both X and Y, $\cap$ is a mathematic symbol denoting the intersection of two elements); and confidence ($|D_X \cap D_Y|/|D_X|$, the ratio of the number of documents containing both X and Y to the number of documents containing X). Thresholds for support and confidence are selected to identify interesting and useful association rules. Concepts that pass the threshold test are used for LBD.

- *Term frequency-inverse document frequency (TF-IDF)*: This is a statistical measure used to evaluate how important a word is to a document in a collection of documents. The importance increases proportionally to the number of times a word appears in the document but is offset by the frequency of the word in the collection of documents. For two concepts, X and Y, the TF-IDF value can be computed as $|D_X \cap D_Y| \times \log(N/|D_X|)$, where N is the size of the document collection. A threshold is selected to identify associated concepts.

- *Z-score*: Z-score mines associations from literature using concept probability distributions. Let V be the set of terms appearing in a document collection. We define $P(x,y) = |D_x \cap D_y|/|D_x|$, where x and y are terms. For two concepts, X and Y, the Z-score value can be computed as: $Z - \text{score}(X,Y) = P(X,Y) - \text{Average}_{x \in V}[P(x,Y)] / Std_{x \in V}[P(x,Y)]$. A threshold is then selected to identify associated concepts.

- *Mutual information measure (MIM)*: This is widely used to quantify dependencies. Assume the size of the document collection to be N, the MIM of two concepts, X and Y, can be computed as: $\text{MIM}(X,Y) = \log_2(|D_X \cap D_Y| \times N)/(|D_X| \times |D_Y|)$. If the concepts are independent, the MIM score equals to zero. Positive MIM score indicates high chance of co-occurrence, while negative MIM score indicates the two concepts are rarely mentioned together.

  For an LBD task, only novel connections are of interest. After pruning pairs known to be correlated, it is desired that the remaining pairs, that is, potential discoveries, are ranked so that researchers can prioritize their explorations. Popular ranking algorithms include the following:

- *Average minimum weight (AMW)*: AMW is based on the assumption of inferring a correlation between starting concept A and target concept C that depends upon how strong the association was in the two associations (A to B and B to C). The overall strength of the association for the inferred association would be no greater than the one associated with the weaker one given by these two associations. The strength of the association can be measured using the metrics used in mining associations described above. For example, assume V is a collection of B concepts and we adopt MIM as the measurement of the strength, the weight of discovery pair (A,C) is computed as: $\text{AMW}(A,C) = \text{Average}_{B \in V} \{ \min [\text{MIM}(A,B), \text{MIM}(B,C)] \}.$

- *Linking term count with average minimum weight (LTC-AMW)*: An alternative assumption could be that the number of B concepts that connect target concept C to A is the main indication of a strong correlation. With this assumption, the association can be ranked according to the number of B concepts. LTC-AMW makes this assumption and ranks the discovery pairs according to the number of B concepts. In case of tie, AMW can then be utilized to prioritize.

14.2.5 Ontology and Semantic Web Techniques

To facilitate the use of computational techniques, associations between domain concepts or entities must be recorded in more subjective ways. Ontological representation has provided a means of connecting concepts or facts from various specialized domains. Meanwhile, the advance in semantic web technologies has produced a set of recommendations for standardizing and manipulating information and knowledge including the Resource Description Framework (RDF), RDF Schema, the Web Ontology Language (OWL), Simple Knowledge Organization System (SKOS), and SPARQL (a query language for RDF graphs). In the biomedical domain, the existence of a variety of software tools such as Protégé has accelerated the development and use of ontology. The online tools and web portal developed by the National Center for Biomedical Ontology (NCBO) provide additional powerful resources to enable scientists to create, disseminate, and manage and analyze biomedical data and information using ontologies and associated resources. An increasing number of biomedical ontological resources such as gene ontology (GO) are now available in OWL format. LBD can be enhanced by the use of ontologies for inferring associations. The Unified Medical Language System (UMLS) which gathers concepts and entities from over 160 sources provides a single standardized format for accessing information across the fragmented biomedical literature. Tools available for detecting UMLS concepts (e.g., MetaMap) and establishing associations among the concepts (e.g., SemRep) have greatly accelerated the discovery pace in LBD.

14.2.6 Network Analysis and Visualization

Literature networks are an intuitive representation of the associations mined from literature. Typically the edges are linked to the underlying literature that connects two concepts, whereas the nodes are linked to databases with additional information about the concepts. A particular useful technology that is used for enrichment of literature network with additional scientific data is the semantic web technology. For example, OpenPhacts is an initiative focused at accelerating drug discovery by connecting clinical, biological and chemical data to pharmacological entities. Literature networks allow mapping of links between two concepts into a space in which multiple links between concepts can be visualized. This has the advantage that also indirect links between concepts become apparent, which can give insight into for instance new relations. Moreover, network-based techniques can be used to detect clusters or network motifs.

14.3 LBD TOOLS/RESOURCES AND ILLUSTRATION

One significant literature source for LBD in the biomedical domain is MEDLINE, the U.S. National Library of Medicine (NLM) premier bibliographic database that contains more than 22 million references to journal articles in life sciences with a concentration on biomedicine. Each reference is indexed with NLM Medical Subject Headings (MeSH). The database can be queried using PubMed, which is a tool providing free access to MEDLINE. Table 14.1 shows some of the LBD tools built on the top of MEDLINE to support literature discovery. One notable resource is Semantic MEDLINE. In the following, we describe Semantic MEDLINE in depth as well as multiple studies using Semantic MEDLINE database to support knowledge discovery.

Semantic MEDLINE integrates information retrieval, advanced NLP, automatic summarization, and visualization into a single web portal. The application is intended to help manage the results of PubMed searches by condensing core semantic content in the citations retrieved. Output is presented as a connected graph of semantic relations, with links to the original MEDLINE citations. The ability to connect salient information across documents helps users keep up with the research literature and discover

TABLE 14.1   List of Literature-Based Discovery Systems

System	Description	URL
ArrowSmith (Smalheiser and Swanson 1998)	Arrowsmith allows users to identify biologically meaningful links between any two sets of articles A and C in PubMed, even when these share no articles or authors in common and represent disparate topics or disciplines.	http://arrowsmith. psych.uic.edu/ arrowsmith_uic/
Chilibot (Chen and Sharp 2004)	Chilibot distills scientific relationships from knowledge available throughout a wide range of biological domains and presents these in a content-rich graphical format, thus integrating general biomedical knowledge with the specialized knowledge and interests of the user.	http://www. chilibot.net/
BITOLA (Hristovski et al. 2001)	An interactive literature-based biomedical discovery support system. The purpose of the system is to help the biomedical researchers make new discoveries by discovering potentially new relations between biomedical concepts.	http://ibmi3. mf.uni-lj.si/ bitola/
LAITOR (Barbosa-Silva et al. 2010)	A text mining system that analyses co-occurrences of bioentities, biointeractions, and other biological terms in MEDLINE abstracts. The method accounts for the position of the co-occurring terms within sentences or abstracts.	http://sourceforge. net/projects/ laitor/
LitInspector (Frisch et al. 2009)	LitInspector lets you analyze signal transduction pathways, diseases, and tissue associations in a snap.	http://www. litinspector.org/
PubNet (Douglas et al. 2005)	A web-based tool that extracts several types of relationships returned by PubMed queries and maps them into networks, allowing for graphical visualization, textual navigation, and topological analysis. PubNet supports the creation of complex networks derived from the contents of individual citations, such as genes, proteins, Protein Data Bank (PDB) IDs, Medical Subject Headings (MeSH) terms, and authors.	http://pubnet. gersteinlab.org/
Semantic MEDLINE (Kilicoglu et al. 2008; 2012)	Semantic MEDLINE is a Web application that summarizes MEDLINE citations returned by a PubMed search. Natural language processing is used to extract semantic predications from titles and abstracts. The predications are presented in a graph that has links to the MEDLINE text processed.	http://skr3.nlm. nih.gov/ SemMed/index. html

connections which might otherwise go unnoticed. Figure 14.2 shows the overall system architecture of Semantic MEDLINE and a screenshot of Semantic MEDLINE. Multiple studies have demonstrated the ability of semantic predication in facilitating literature discovery (Deftereos et al. 2011). For example, Hristovski et al. (2013) used semantic predications

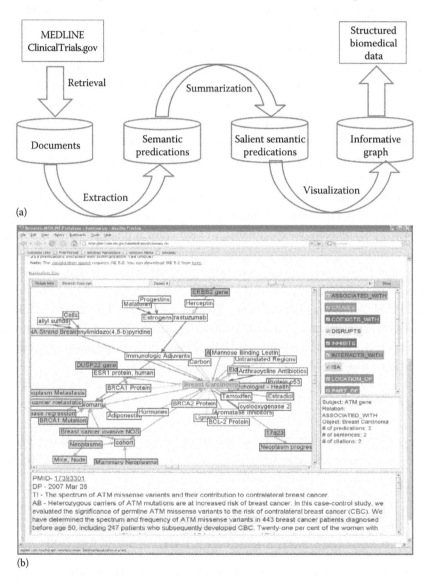

(a)

(b)

FIGURE 14.2 (a) Overview of the system architecture of Semantic MEDLINE. (b) Screenshot of Semantic MEDLINE (http://skr3.nlm.nih.gov/SemMed/ SemMedQuickTour/SemMedQuickTour_player.html).

extracted by two NLP systems to develop a LBD system which produces a smaller number of false positive and facilitates user evaluation process of potentially new discoveries.

The backend of Semantic MEDLINE is Semantic MEDLINE Database (SemMedDB), a repository of semantic predications (subject–predicate–object triples) extracted by an association extraction system, SemRep. Comparing to co-occurrence-based LBD, the use of SemMedDB keeps only meaningful predications. We review two studies where one uses Semantic MEDLINE for discovering drug–drug interactions and the other focuses on network analysis formed by semantic predication pairs.

Using semantic predications to uncover drug–drug interactions in clinical data (Zhang et al. 2014): Drug–drug interactions (DDIs) are a serious concern in clinical practice, as physicians strive to provide the highest quality and safety in patient care. DDI alerts are commonly implemented for in CPOE (computerized physician order entry) systems. However, some DDIs result from combinations or various mechanistic pathways that are not widely known. DDIs can be identified through several approaches, including in vitro pharmacology experiments, in vivo clinical pharmacology studies, and pharmacoepidemiology studies. However, these methods are limited by the need to focus on a small set of target proteins and drugs and are, therefore, slow to elucidate an exhaustive set of DDIs while new drugs are continually added into the pharmacopeia. Because they depend on these methods of DDI discovery and anecdotal clinical reporting, current DDI databases do not include all of the potential DDIs. Zhang et al. proposed an approach which utilizes SemMedDB for discovering potential DDIs.

Specifically, for a given list of drugs, semantic predications of the following are retrieved:

- Predications describing an influence between a drug and a gene—drug–gene or gene–drug pairs with predicate types as INHIBITS, INTERACTS_WITH, and STIMULATES.

- Predications describing the biological functions of genes—gene–function pairs with predicate types including AFFECTS, AUGMENT, CAUSES, DISRUPTS, INHIBITS, and PREDISPOSES.

- Predications describing the biological functions of drugs—drug–function pairs having a drug as the subject and a function as the object where the function refers to all descendants of the biological function semantic type in the UMLS semantic-type hierarchy.

Two DDI discovery pathways are proposed. One is drug1-gene-drug2 (DGD) pathway. Potential DDI candidates are generated by identifying (drug1, gene) and (gene, drug2) pairs in the retrieved semantic predications. The other pathway is drug1-gene1-function-gene2-drug2 (DGFGD) pathway. In this scenario, there is no gene commonly shared by the two drugs but through a common biological function, the two drugs are associated. To ensure the function is not an established effect of either drug, the pathway only retains the drug pairs through function if there are no predications linking drug1 or drug2 to function directly.

An evaluation demonstrates that the approach found 19 known and 62 unknown DDIs for 22 patients randomly selected. For example, the interaction of Lisinopril, an ACE inhibitor commonly prescribed for hypertension, and the antidepressant sertraline can potentially increase the likelihood and possibly the severity of psoriasis.

Network analysis of a drug-disease-gene predication network (Zhang et al. 2014): For a given set of predications, a predication network can be formed by treating entities as nodes, and predications as edges. However, such network can be noisy due to (1) predications extracted may not be accurate due to the limitation of the extraction tool and (2) the association may not be scientifically correct. Network-based computational approaches can be utilized to analyze the network by decomposing them into small subnetworks, called *network motifs*. Network motifs are statistically significant recurring structural patterns found more often in real networks than would be expected in random networks with same network topologies. They are the smallest basic functional and evolutionarily conserved units in biological networks. The hypothesis is that network motifs of a network are the significant sub-patterns that represent the backbone of the network, which serves as the focused portion. These network motifs could also form large aggregated modules that perform specific functions by forming associations among a large number of network motifs.

The network analysis was performed on a drug-disease-gene prediction network formed by two of the three entities, drug, disease, and gene where drugs are from a collection of FDA-approved drug entities in DrugBank. The network consists of 84,317 associations among 7,234 entities (including drugs, diseases, and genes), which is too complex for a direct visualization. Through network motif analysis, five significant three-node network motifs are identified, which have strong biological meanings and could suggest scientists' future directions in their research field (Figure 14.3).

(a)　　　　　(b)　　　　　(c)　　　　　(d)　　　　　(e)

FIGURE 14.3   Network motifs detected through motif analysis. Triangle—disease, cycle—gene, and square—drug. (a–e) Five significant three-node network motifs identified through network motif analysis.

Three case studies were performed:

- *Prioritize disease genes*: One of the motifs is disease1-gene-disease2-disease1, which indicates two diseases that are associated with each other are also associated with one common disease gene. By limiting the network to this motif, we can prioritize disease genes.

- *Inference of disease relationships*: Another significant motif is disease1-gene-disease2, where two diseases are associated with a common gene but the association of disease1 and disease2 is not a requirement. Based on the "guilt by association" rule—diseases similar to each other are more likely to be affected by the same genes/pathways, two diseases involved are more likely to be similar/associated than other diseases.

- *Drug repositioning*: Limiting the network to disease1-drug-disease2-disease1, where two diseases associated with each other, can be the targets for the same drug. This motif indicates similar diseases can be treated by same drugs, allowing us to make hypotheses for drugs repositioning purpose.

14.4 STEP-BY-STEP TUTORIAL ON SIMPLE LBD USING SEMANTIC MEDLINE

All semantic predications for Semantic MEDLINE were stored in a relational database called SemMedDB. This repository of the predications is publically available and can be downloaded from http://skr3.nlm.nih.gov/SemMedDB/. To use this database, the user should install MySQL and then import the repository. Installing MySQL is pretty simple and straightforward for different platforms such as Microsoft Windows, Mac OS, and Linux. We assume the reader has installed MySQL on his/her system and is familiar with SQL commands. After installing MySQL, first we need to create a database. In the shell script enter

```
mysql -h localhost -u USERNAME -p
```

and then enter your password. This command enters you to MySQL shell, which allows you to run SQL commands. If your username and password is correct, MySQL prints a welcome message and it is ready to run your commands. To create a database enter this command:

```
Create database DATABASE_NAME;
```

If the command was executed correctly, MySQL prints this message *Query OK, 1 row affected (0.00 sec)*. Now, you should dump SemMedDB into this database. There are two options to download SemMedDB: (1) the whole database and (2) each table separately. After downloading the database, either way, first exit MySQL shell via typing *exit* command and then enter the following command to import SemMedDB into the new database:

```
mysql -u USERNAME -p -h localhost DATABASE_NAME <
DOWNLOADED_FILE.sql
```

This operation takes time. After importing the database, go back to MySQL shell. Enter "use DATABASE_NAME;". Type "show tables;" to view all the tables in the database. The database contains 8 tables. One of the tables called PREDICATION_AGGREGATE contains all the data that we need to generate LBDs. Some of fields in this table are unique identifier of PubMed Citation (PMID), name of subject (s_name), semantic type of subject (s_type), name of object (o_name), semantic type of object (o_type), type of predicate, and unique identifier to the sentence (SID) that contains the predication. This table allows us to generate both open and closed LBD.

14.4.1 Retrieving Open Discoveries

Assume we want to generate a list of open LBDs between drug (start concept) and disease (target concept), with gene as linking concept. Three steps should be taken for this retrieval:

```
#First retrieving semantic predications between drug
and gene
SELECT      PMID, s_name, o_name, predicate, SID
FROM PREDICATION_AGGREGATE
WHERE s_type='phsu' AND o_type='gngm'  AND predicate
NOT Like 'neg_%'
as T1;
```

```
#Second retrieving semantic predications between gene
and disease
SELECT PMID, s_name, o_name, predicate, SID
FROM PREDICATION_AGGREGATE
WHERE s_type='gngm' AND o_type='dsyn' AND predicate
NOT Like 'neg_%'
as T2;


#Joining these two tables
SELECT T1.s_ name, T2.o_ name
FROM T1, T2
WHERE T1.o_name=T2.s_name AND T1.PMID <> T2.PMID
```

which ***phsu*** is semantic type for pharmacologic substance, ***gngm*** for gene or genome, and ***dsyn*** for disease or syndrome. The negated predications extracted by SemRep are removed by this condition: ***predicate NOT Like 'neg_%'***. Here is the whole query:

```
SELECT T1.s_ name, T2.o_ name
FROM
(SELECT PMID, s_name, o_name, predicate, SID
FROM PREDICATION_AGGREGATE
WHERE s_type ='phsu' AND o_type='gngm'  AND predicate
NOT Like 'neg_%') as T1 ,

(SELECT PMID, s_name, o_name, predicate, SID
FROM PREDICATION_AGGREGATE
WHERE s_type='gngm' AND o_type='dsyn' AND predicate
NOT Like 'neg_%') as T2

WHERE T1.o_name=T2.s_name AND T1.PMID <> T2.PMID
```

To retrieve the sentences, which the semantic predications are extracted from, we should join another table, called SENTENCE, to the query. The following steps retrieve the sentences along LBDs:

```
#Joining SENTENCE and T1

SELECT *
FROM T1 and SENTENCE
WHERE SENTENCE.SID = T1.SID
as T3;
```

```
# Joining SENTENCE and T2
SELECT *
FROM T2 and SENTENCE
WHERE SENTENCE.SID = T2.SID
as T4;


#Joining these two tables
SELECT T3.s_ name, T3.sentence, T4.o_ name,
T4.sentence
FROM T3, T4
WHERE T3.o_name=T4.s_name    AND    T1.PMID <> T2.PMID
```

14.4.2 Retrieving Closed Discoveries

With a small modification, the above queries can be used for closed discovery. The following query retrieves all semantic predications between **Flecainide** and **Heart Failure**:

```
SELECT        T1.PMID, T1.o_name, T2.PMID
FROM
(SELECT PMID, s_name, o_name, predicate
FROM   PREDICATION_AGGREGATE
WHERE s_name= 'Flecainide' AND o_type='gngm' AND
predicate NOT Like 'neg_%') as T1 ,
(SELECT PMID, s_name, o_name, predicate, SID
FROM PREDICATION_AGGREGATE
WHERE s_type='gngm' AND o_name= 'Heart Failure' AND
predicate NOT Like 'neg_%') as T2

WHERE   T1.o_name=T2.s_name AND T1.PMID <> T2.PMID
```

This query returns 169 rows, which can be used to evaluate the hypothesis of existence of a relationship between these two entities.

REFERENCES

Barbosa-Silva, A., T. G. Soldatos et al. (2010). LAITOR--literature assistant for identification of terms co-occurrences and relationships. *BMC Bioinformatics* **11**: 70.

Chen, H. and B. M. Sharp (2004). Content-rich biological network constructed by mining PubMed abstracts. *BMC Bioinformatics* **5**: 147.

Deftereos, S. N., C. Andronis et al. (2011). Drug repurposing and adverse event prediction using high-throughput literature analysis. *Wiley Interdisciplinary Reviews Systems Biology and Medicine* **3**(3): 323–334.

Douglas, S. M., G. T. Montelione et al. (2005). PubNet: A flexible system for visualizing literature derived networks. *Genome Biology* **6**(9): R80.

Frisch, M., B. Klocke et al. (2009). LitInspector: Literature and signal transduction pathway mining in PubMed abstracts. *Nucleic Acids Research* **37**(Web Server issue): W135–W140.

Hristovski, D., T. Rindflesch et al. (2013). Using literature-based discovery to identify novel therapeutic approaches. *Cardiovascular & Hematological Agents in Medicinal Chemistry* **11**(1): 14–24.

Hristovski, D., J. Stare et al. (2001). Supporting discovery in medicine by association rule mining in Medline and UMLS. *Studies in Health Technology and Informatics* **84**(Pt 2): 1344–1348.

Kilicoglu, H., M. Fiszman et al. (2008). Semantic MEDLINE: A web application for managing the results of PubMed Searches. *Proceedings of the 3rd International Symposium for Semantic Mining in Biomedicine* (SMBM 2008), Finland, pp. 69–76.

Kilicoglu, H., D. Shin et al. (2012). SemMedDB: A pubMed-scale repository of biomedical semantic predications. *Bioinformatics* **28**(23): 3158–3160.

Smalheiser, N. R. and D. R. Swanson (1998). Using ARROWSMITH: A computer-assisted approach to formulating and assessing scientific hypotheses. *Computer Methods and Programs in Biomedicine* **57**(3): 149–153.

Yetisgen-Yildiz, M. and W. Pratt (2009). A new evaluation methodology for literature-based discovery systems. *Journal of Biomedical Informatics* **42**(4): 633–643.

Zhang, R., M. J. Cairelli et al. (2014). Using semantic predications to uncover drug-drug interactions in clinical data. *Journal of Biomedical Informatics* **49**: 134–147.

Zhang, Y., C. Tao et al. (2014). Network-based analysis reveals distinct association patterns in a semantic MEDLINE-based drug-disease-gene network. *Journal of Biomedical Semantics* **5**: 33.

## CHAPTER 15

Mitigating High Dimensionality in Big Data Analysis

Deendayal Dinakarpandian

CONTENTS

15.1 INTRODUCTION

The Greeks had a very simple explanation of life. Both mental and physical health depended on the proper balance between four humors—sanguine, phlegmatic, melancholic, and choleric related to blood, phlegm, black bile, and yellow bile, respectively. We now know that this is too simple a model to make accurate predictions regarding health. For example, if we assume that each of these humors can take on high or low values, there can be only 2^4 or 16 states of health possible. Over a 100 Nobel Prizes later, we feel we are in a much better position to measure the state of the body. While it once took days of work to isolate a single protein from bovine liver, it is becoming increasingly feasible to affordably measure thousands of components in minutes. The components may be metabolites, proteins, DNA, or RNA. We can also measure the extent of chemical transformations of these, for example, glycosylation, phosphorylation, acetylation, or methylation.

There has been a concomitant development in computational and statistical methods to deal with the deluge of measured data. Computational methods have facilitated scalable storage, querying, and comparison of data while statistical methods make it possible to distinguish dependencies from coincidental correlations. This chapter focuses on a few key concepts that have become the *sine qua none* underlying statistical modeling and machine learning. In particular, it helps to understand why being data rich doesn't necessarily translate to reliable knowledge. Rather, we now have a poverty of riches that is plagued by false positives.

The tenor of the chapter is empirical rather than theoretical, in line with the applied focus of the book. Section 15.2 gives a conceptual explanation of the problem posed by having rich data. Section 15.3 presents possible strategies in tackling the problem. The chapter ends with a section that illustrates how some of these strategies can be used in R.

15.2 POVERTY OF RICHES

This section presents various ways in which the curse of dimensionality can undermine drawing reliable conclusions from the analysis of rich data.

Comparing single means: Consider the case of comparing the value of a single variable between two groups of subjects. For example, one may measure the blood level of a protein that is suspected to be elevated in patients with a certain type of cancer. The mean levels the protein in the two groups may be compared to determine if they are significantly different by employing a *t*-test. In another case, the frequencies of an SNP in two different groups may be compared to determine if they are significantly different. Fisher's exact test or a chi-squared test may be used in this case. In both of these examples, there is a single variable being measured—a real number in the first case and a Boolean variable (YES = suspected SNP variant is present; NO = suspected SNP variant is absent) in the second in several observations (number of subjects).

Comparing several means: In contrast to the relatively straightforward problem of comparing just one mean with another, the situation becomes more involved when the ratio between the number of variables and the number of observations increases. In the above examples, this situation arises when a whole metabolic profile is measured in the blood instead of a single molecule or when a large number of SNPs are compared between two groups. In effect, the *p*-values calculated in this case often represent only a lower bound for the actual *p*-value. For example, if the *p*-value for one of 100 SNPs compared between two groups is found to be 0.01, this could actually be much higher (see Bonferroni correction below for explanation).

Multiple questions: In fact, this problem is not restricted to *omic* analyses. It arises in a wide range of experiments such as questionnaires that include a score for each of several questions. Again, if the *p*-value for the difference in score of one of 10 questions between two groups of participants is 0.01, the actual *p*-value could be much considerably higher.

Multiple types of explorations: In addition to the measurement of multiple variables that are of the same type (e.g., bacteria in microbiome analysis, RNA in next-generation sequencing), this problem also arises when several different types of experiments are performed in search of the answer to a question. For example, if one searches for evidence of the presence of a new virus, there may be multiple types of serological or cell culture assays carried out. Once again, getting a significantly higher titer on one of the assays could be just a coincidence.

Regression analysis: Consider the problem of regression analysis where the variables that describe the data are the input and the output is a numerical value. An example is the prediction of survival time given a combination of clinical and molecular data. Irrespective of the actual mathematical technique used for regression, the predictive model is built based on a subset of data referred to as the *training set*. The error between predicted values and the actual values of the output is commonly used as an estimate of the accuracy of the regression equation. *If the number of variables is greater than the number of observation points in the training set, the error will effectively be zero.* To understand this, consider the extreme case of trying to deduce a regression line from a single observation (point). For example, this could be the blood level of a suspected marker from a single patient. In this case, the number of observations is the same as the number of variables. We intuitively know that it is impossible to derive any general conclusion from a single observation. The mathematical explanation is that an infinite number of lines can pass through a single point.

There is no way of knowing which is the correct line that should be used for prediction. On the other hand, if there are two observations available, then there is only one line that will pass exactly through both points.

Similarly, consider the case of two observations, but now using two variables for prediction. For example, data from two patients on the blood levels of two different markers may used to make a prediction. In this case, the regression function corresponds to a plane because it exists in three-dimensional space—two dimensions for the input, and one for the output. However, since the number of observations is equal to the number of variables, there is an infinite number of predictive models that have an error of zero. The mathematical explanation is that an infinite number of planes can pass through a single line. Essentially, if the number of observations is not greater than the number of variables used as input, the training set ceases to be a useful guide to identify the relationship between input and output, and one cannot learn anything from the available data. The same concept holds when one may have data from a hundred patients but a thousand SNPs are used to quantify underlying relationships. In this case, an infinite number of hyperplanes can map perfectly to the training data, yielding a perfect but spurious prediction.

Further, it is important to note that the problem persists even when the number of variables is less than the number of observations, but is still relatively large compared to the number of observations. Though there is no longer an infinite number of possible regression equations with perfect prediction, and there may a unique best model for a given set of training data, the nature of the model may vary considerably depending on the subset chosen to be the training data. This phenomenon is referred to as *model instability*, since the predictive equation changes considerably from one training set to the next. For example, completely different sets of SNPs may be considered important by different groups of researchers who have used different training sets (subjects). In effect, the flexibility afforded by the large number of variables or dimensions influences the learning process to effectively *memorize* the training data and therefore be unable to distinguish the underlying general trend from random (noisy) fluctuations. This phenomenon is commonly referred to as *overfitting*.

Classification: In classification, we seek to allocate each observation to a category, for example, normal, borderline, or disease. In this case, predictive error may be quantified by measures like specificity, sensitivity, precision (for two categories), or by weighted average of accuracies over multiple categories. Analogous to the potential problems with regression, the same issue of spurious prediction caused by rich data occurs in classification problems. The line, plane, or hyperplane in this case doesn't in itself represent the output; rather, it divides the classification space into the different categories. Other than this difference in interpreting the input–output space, all the considerations are similar. If there are only two observations for two variables used as predictors, there are an infinite number of ways to perfectly classify the data. And if the number of variables is less than the number of observations but still considerable, spurious models are likely to be generated. The prediction will appear to be near perfect on a given training data set, but will have an intolerable amount of error on test sets.

Curse of dimensionality: To summarize, a tempting solution to answering scientific questions is to acquire expensive instrumentation that can give a rich description of organisms or biological conditions by measuring large numbers of variables simultaneously. However, it is fallacious to assume that this makes prediction easier.

15.3 DATA ANALYSIS OUTLINE

A variety of representative approaches to tacking the curse of dimensionality are summarized in this section. A naïve approach might be simply to increase the number of observations. This line of thinking underlies power analysis used by statisticians to reduce the probability of false negatives. However, the problem here is the need to reduce false positives. Increasing the number of observations as a possible solution is often infeasible for economic reasons (cost of recruiting subjects or obtaining biological samples) or for scientific reasons (rare conditions). More importantly, the number of observations required for reliable prediction rises exponentially as a function of the *richness* (number of measured predictors or variables) of data. This implies that the dimensionality of predictive problems will dwarf the number of possible observations in the large majority of cases; there will never be enough data. The following types of strategies are used to mitigate the problem of spurious findings. Broadly speaking, there are two choices—reduce the number of dimensions by judicious screening or perform a correction to compensate for the (large) number of dimensions used. In practice, a suitable combination of the following strategies is recommended for any given project.

Using prior knowledge to reduce dimensions: A common temptation is to carry out statistical analysis or computational prediction by using as many variables as possible in a *multi-everything* approach that is apparently comprehensive. The naïve line of thinking is that nothing will be missed by using this approach. As discussed in the previous section, this attempt to minimize false negatives is more likely to result in false positives or erroneous models. Therefore, it is better to use scientific knowledge of the problem to eliminate irrelevant variables and thus increase the ratio of observations to predictors. Alternatively, variables could be grouped into a smaller number of dummy variables. One example of this is to choose to use a fewer number of gene ontology terms instead of individual genes in prediction.

Using properties of data distribution to reduce dimensions: Properties of the data like correlation between dimensions and asymmetry of how the data are distributed may be exploited to reduce the total number of variables. For example, if a subset of the variables exhibits correlation with each other, one of the variables may be chosen as a representative or a virtual composite variable may be used. Alternatively, a new (smaller) set of variables may be created by using a non-redundant (orthogonal) version of the original dimensions. Eigenanalysis of the data may be performed to determine which directions show the greatest variance—as in principal component analysis. The original data may be mapped to a new (smaller) space where each *variable* is a principal component.

Using exploratory modeling to reduce number of dimensions: Irrespective of the prediction method used, it is possible to use systematic approaches to find a minimal subset of the original variables that represents a good compromise between the spurious nearly perfect prediction with everything included and a more reliable prediction with apparently higher error. These may be performed by gradually increasing the number of variables considered (forward subset selection), decreasing the number of variables considered (backward subset selection), or a combination (hybrid). The key consideration is to use the smallest number of variables that lowers the prediction error to an acceptable amount.

Penalizing schemes for complex models: A model is considered complex if it has many parameters or coefficients. While higher complexity is desirable for accurate predictions where the underlying mechanisms are complicated, too complex a model can cause higher error by overfitting. In addition to the number of coefficients in a model, the magnitude of coefficients is also a measure of complexity. Given two models that give similar performance, it is better to prefer the one with lower complexity. This principle is often referred to as *Ockham's razor*. One way to find the appropriate level of complexity is to compare solutions by adding a penalty that is proportional to the complexity. A couple of methods used in linear regression are least absolute shrinkage and selection operator (LASSO) (Tibshirani 1996) and ridge regression (Hoerl and Kennard 1970). LASSO penalizes additive models by the magnitude of the coefficients. Without going into details of the underlying mathematics, LASSO tends to discard irrelevant variables or dimensions. Ridge regression penalizes additive models by the squares of the coefficients. While this does not discard any of the dimensions, it helps to reduce the importance of less influential variables in prediction.

Correction for multiple testing: As mentioned before, the availability of richly described data can be counter-productive in generating many false positive results. The *p*-values need to be corrected (increased) to differentiate between false and true positives. Consider two experiments. In one experiment, the level of expression of a single gene is compared between two groups of subjects. The *p*-value corresponding to the difference is .0001. This implies that the probability that there is no difference between the two groups is at most .0001. Contrast this with a second experiment, where the levels of expression of a one hundred genes are compared between two groups. Even though this sounds like a single experiment, in reality this is a set of hundred separate experiments that just happen to be performed together. If one of the genes shows a difference with a *p*-value of .0001, this corresponds to finding such a result once in 100 attempts. Therefore, one way to correct each of the *p*-values is to multiply them by the number of *actual* experiments, yielding .001 as the corrected value for a *raw* value of .0001. This is referred to as the *Bonferroni correction* (Bonferroni 1936). Another way of viewing this is to adjust the *p*-value thresholds for statistical significance. For example, the equivalent probability of a coincidental finding that occurs 5% of the time in a single experiment is 0.05/100 if observed among 100 experiments. While the Bonferroni correction does a good job of eliminating coincidental findings, it can also throw the baby out with bath water by increasing false negatives. In practice, the number of experiments is so large (thousands of gene comparisons, for example) that the required *p*-value threshold for significance may be unrealistically low. The reason is because the Bonferroni correction assumes that variables do not interact with each other. This notion of independence is rarely true in biomedical research. By definition, the concentrations of protein or RNA molecules are often dependent on each other. In other words, if a hundred genes are studied, the number of independent subgroups within them is usually smaller. This implies that the *true p*-value lies somewhere between the raw values and the value obtained by multiplying them with the total number of comparisons made.

An alternative to the aggressive Bonferroni method for multiple testing corrections is to answer a slightly different question. Instead of asking the Bonferroni question of "What is the probability that there is at least one spurious/coincidental difference in this group of experiments?" one tries to answer the question "What proportion of the observed differences is not really a difference but just a spurious/coincidental difference?" For example, one could ask, how many genes showing a difference with a *p*-value of less than or equal to 0.05 are in reality not different? This is referred to as the *false discovery rate* (FDR). In other words, the goal is to find the appropriate threshold that corresponds with a desired FDR. This can be done empirically by permutation testing—shuffling the observed values and creating a frequency histogram of the resulting coincidental differences between two groups. This can then be compared with the actual values to estimate the FDR. For example, if there are 40 genes that show a difference in expression greater than 2 in the real experiment, and there are 4 genes that show at least a difference of 2 in the real experiment, then the corresponding FDR for a threshold of 2 is 4/40 or 10%. So if one desires a FDR of 5%, then we might slide the threshold and find that a cutoff difference value of 3 yields 20 genes in the actual set versus 1 in the spurious set—corresponding to 1/20 or the desired FDR of 5%. As an alternative to permutation testing, the threshold value for a desired FDR can also be determined by sorting the original *p*-values and following the Benjamini–Hochberg procedure to obtain equivalent results (Benjamini and Hochberg 1995).

Importance of cross-validation: A variety of strategies for dealing with the curse of dimensionality has been outlined above. A universal method for evaluating performance, irrespective of the actual method used to build a predictive model, is cross-validation. A common error is to use the entire data available to build a model and then end up using the same data to evaluate the performance. This is likely to result in a highly optimistic estimate of performance. A slightly better approach is to divide the data into two parts, using one for training and the other for validation. However, it is possible that the division is fortuitous such that the testing data are very similar to the training data. An even better method is 10-fold cross-validation, where the data are randomly divided into 10 subsets followed by rounds of evaluation. In each round, 9 of the subsets are pooled together for training, with a single subset being used to evaluate the performance. If the number of observations is limited, leave-one-out validation can be used where each subset is essentially a single observation. In either case, every observation ends up being used for both training and testing to give a realistic average estimate of prediction error on a new test observation. Cross-validation in itself is not a solution for the curse of dimensionality but it yields an objective measure of predictive performance and can be used to rank models.

15.4 STEP-BY-STEP TUTORIAL

In this section, a brief tutorial on how to correct for multiple testing in R is presented.

Consider two data sets exp1 and exp2, which represent the expression values of a single gene in two groups of 50 subjects each. Assume that the normalized mean value in both groups is zero, with a standard deviation of 1. The following commands may be used to create the corresponding data sets:

```
> exp1=rnorm(50, mean=0,sd=1)
> exp1
```

```
[1]  -0.57068957   0.04736721 -1.24841657 -1.44353435   0.49662178
[6]   2.60968270 -0.96959959 -1.13274380 -0.33420912 -0.55439927
[11] -0.60649030 -0.46635946   0.19279477 -1.29596627   0.45703230
[16] -0.86291438 -1.65985004   0.41464152 -1.30537486 -0.40097109
[21] -0.04646163 -1.36372776 -0.91189955   0.20931483   1.17841029
[26] -1.23847239 -1.23736365 -0.16658649 -0.16345373   0.21434718
[31]  0.97866365   0.30745350 -0.26211568 -0.29154925   0.65174597
[36]  0.87553386   0.88960715   0.04319597   0.98085568 -2.20208429
[41] -0.15386520   0.58222503   0.46074241   0.21359734   0.81942712
[46] -1.64504171   0.81400012   0.56407784   0.94932426   1.08691828
> exp2=rnorm(50, mean=0,sd=1)
```

Since the values are normally distributed, a *t*-test may be performed to see if the mean values in the two experiments are different. The relevant command in R is given below. Even though the means of the two sets of values appear to be different, the high *p*-value of .4 suggests that there is no significant difference in the mean value of this gene between the two groups; there is a 40% chance that both sets of values are samples of the same distribution.

```
> t.test(exp1,exp2)
     Welch Two Sample t-test
data:  exp1 and exp2
t = -0.7911, df = 97.621, p-value = 0.4308
alternative hypothesis: true difference in means is
not equal to 0
95 percent confidence interval:
-0.5446399  0.2341746
sample estimates:
 mean of x   mean of y
-0.12993119  0.02530144
```

The above commands may also be carried out by reading in a file containing a single column of measurements of a single gene under a particular condition. A wide variety of formats can be read by R.

Now consider the situation of measuring the expression level of 100 genes in two groups of 50 subjects each. In this case, a real data set would be formatted as 50 rows of 100 values each. For purposes of illustration, synthetic data sets are created and used below.

The data for the first group (normal distribution with mean value = 0, standard deviation = 1) are created as an array of 50 rows (subjects) and 100 columns (genes).

```
> Exp1with50subjects100genes<-array(rnorm(5000,0,1), c(50,100))
```

A partial view of the data showing the values of the first 5 genes of the first five subjects is shown below.

```
> Exp1with50subjects100genes[1:5,1:5]
             [,1]          [,2]         [,3]         [,4]         [,5]
[1,]   1.417706924   0.6755287 -1.1890533   1.1397988   0.2501593
[2,]   0.004892736  -1.5852356 -0.8496448   0.9739892  -0.1589698
[3,]  -0.997550929   0.3602879 -0.8737415   1.0237264  -0.3268001
[4,]   0.237741476   0.1917299  1.2006769   1.4636745  -0.5755778
[5,]  -0.846503962  -0.6692146 -0.6169926  -0.3442893   0.7648831
```

The values for the first two genes can be deliberately altered to have mean values of 2 and 0.6, respectively.

```
> differentGene1 = array(rnorm(50,2,1))
> differentGene2 = array(rnorm(50,0.6,1))
> Exp1with50subjects100genes[1:5,1:5]
             [,1]          [,2]         [,3]         [,4]         [,5]
[1,]   1.417706924   0.6755287 -1.1890533   1.1397988   0.2501593
[2,]   0.004892736  -1.5852356 -0.8496448   0.9739892  -0.1589698
[3,]  -0.997550929   0.3602879 -0.8737415   1.0237264  -0.3268001
[4,]   0.237741476   0.1917299  1.2006769   1.4636745  -0.5755778
[5,]  -0.846503962  -0.6692146 -0.6169926  -0.3442893   0.7648831
> Exp1with50subjects100genes[,1] = differentGene1
> Exp1with50subjects100genes[,2] = differentGene2
> Exp1with50subjects100genes[,1:5]
            [,1]           [,2]           [,3]          [,4]          [,5]
[1,]   2.1374105  -1.58119132  -1.189053260   1.139798790   0.25015931
[2,]   2.6269713   1.09021729  -0.849644771   0.973989205  -0.15896984
[3,]   3.0281731   0.52403690  -0.873741509   1.023726449  -0.32680006
[4,]   3.1437150  -0.98168296   1.200676903   1.463674516  -0.57557780
[5,]   1.7539795   0.02055405  -0.616992591  -0.344289303   0.76488312
[6,]   1.2765618  -0.29014811  -0.816515383  -0.445786675   0.73574435
[7,]   4.4226439   0.13519754   0.172053880   2.061859613  -0.59618714
[8,]   2.0093378   1.74462761  -0.089958692  -0.478425045  -2.78549389
[9,]   2.3642556   1.63258480  -0.487598270  -2.732398629   1.22911743
[10,]  1.6724412   0.83969361  -0.502998447   0.065667490  -0.31565348
[11,]  1.2369272   0.30434891   0.920655980   1.055611798  -0.45456017
[12,]  2.3038377  -0.56758687   1.115077162   1.134437803   0.06946009
[13,]  2.1306358   1.28862167  -2.393985146  -0.433934763   0.47876340
[14,]  1.2301407   0.73632915  -0.100082003   0.406445274  -0.01973016
[15,]  1.6220515   0.65160743   1.034377840  -1.763653578   0.68346130
[16,]  1.8190719   0.42323948   0.866958981   0.809686626  -0.47866677
[17,]  2.7460363  -0.01443635  -1.715260186  -0.187892145  -0.61911895
[18,]  1.8095900   1.50900408   0.810839357   1.288867130  -0.37689579
[19,]  2.1988674   0.29528122  -0.086798788   0.983140330  -0.26887477
[20,]  1.3043292   1.69655976   0.093611374   0.452242483   0.14640593
[21,]  1.8725796   0.50599179  -1.074072964  -0.306025566   1.47509530
[22,]  2.7421148   1.81896864   2.093938163  -0.941033776  -1.38505453
[23,]  5.5086279   0.68334959  -0.282948018  -0.442854621   0.93116003
[24,]  2.0330395   0.25910465  -1.391611379   1.702541228  -1.15721324
```

```
[25,]  1.6247933 -0.22164555 -0.260075758 -0.205446006  0.85312450
[26,]  1.2317709  1.21296715 -0.037042265 -0.133936305 -1.39644257
[27,]  2.3996270  1.16273939  0.655105362 -1.446995186 -0.76728085
[28,]  1.2956523  0.68987290 -2.104914939 -1.434816121  0.76627595
[29,]  2.1268622  1.62353243 -0.182621578  0.352729492  0.30991314
[30,]  3.0178016  1.85885646 -0.435628448  2.559880444  0.86493823
[31,]  2.4915555  0.43702543  0.267144603  1.588996864  0.37721399
[32,]  2.4907016  1.83148665  0.005977475  1.024481087  0.03755617
[33,]  1.5534359  1.36063142 -0.964298354 -2.246160891 -1.16830203
[34,]  1.4502630  1.15023216 -0.095982013  0.903945360 -1.78614936
[35,]  2.5028584  0.45545447 -0.852866573  0.153783706 -0.30751374
[36,]  1.1671471  0.16062241  0.347215168  0.007958926  0.01619001
[37,]  0.2705531  1.69753007 -1.268761517 -0.310926359 -2.12551675
[38,]  0.6374114 -0.39832964 -1.711567105  0.224159122  0.22434726
[39,]  2.3548472  1.93852588  0.329526906 -1.429820435 -0.25518701
[40,]  1.8328574  1.77574099 -0.706425569 -0.797979554  0.24350870
[41,]  2.5490908  0.06858288  0.458804390 -0.021129068 -0.20909139
[42,]  2.5149380 -0.26304752  0.127733387  0.446516390 -0.76222503
[43,]  2.3675076  0.23149459 -0.335212884  1.253704434 -0.07246676
[44,]  1.8292905 -0.29737483  2.199146595 -0.673142695  0.56156799
[45,]  1.8252365 -0.01701498  0.806919423  0.896151150  2.53778543
[46,]  3.0843245  2.26915560 -0.911682656 -1.207595664 -0.05926692
[47,]  1.5347754  0.24529655 -0.246986436 -0.827229970 -0.07835391
[48,]  3.1281320  0.59786603 -1.163498286  2.415567289 -0.72253147
[49,]  2.2260141 -0.32585954 -0.155022913  1.841579340  1.82398766
[50,]  1.7593475  1.80584131 -1.229134523  1.306847191 -0.12366554
```

A view of the data set now shows that the first two columns (see above) have higher values than the others. The means estimated from the data are seen to be close to the expected values of 2, 0.6, and 0 for the first three columns.

```
> mean(Exp1with50subjects100genes[,1])
[1] 2.125203
> mean(Exp1with50subjects100genes[,2])
[1] 0.6754891
> mean(Exp1with50subjects100genes[,3])
[1] -0.2325444
```

A second data set with all genes having a mean of 0 is created for statistical comparison with the first set. The estimated means for the first two columns, like the others, are close to zero.

```
> Exp2with50subjects100genes<-array(rnorm(5000,0,1),
c(50,100))
> mean(Exp2with50subjects100genes[,1])
[1] -0.1651355
> mean(Exp2with50subjects100genes[,2])
[1] -0.1475769
```

```
> mean(Exp2with50subjects100genes[,3])
[1] -0.2552059
```

The two data sets are compared by performing 100 *t*-tests, one for each gene or dimension.

```
> pvalues = c(1:100)
> p=numeric(0)
> for (i in 1:100)
+ pvalues[i] = t.test(Exp1with50subjects100genes[,i],
Exp2with50subjects100genes[,i])$p.value
> pvalues
```

The resulting 100 *p*-values are shown below. As expected, the first two *p*-values are low. However, we find that there are several spurious *p*-values—genes 4, 42, 53, 80, 84, 99—that are lower than the traditional threshold of 0.05. These indicate false positives.

```
 [1] 5.415589e-22 4.529549e-05 9.097079e-01 1.486974e-02 9.054391e-01
 [6] 7.098275e-01 4.704718e-01 8.877928e-01 9.196552e-01 9.164268e-01
[11] 1.915345e-01 8.739235e-01 1.242989e-01 9.936854e-01 7.761322e-01
[16] 6.143852e-01 2.527180e-01 4.983262e-01 5.364489e-01 3.473893e-01
[21] 7.880777e-02 3.001068e-01 9.102016e-01 4.904164e-01 3.241277e-01
[26] 6.777727e-01 1.861560e-01 9.631817e-01 8.152621e-01 9.847419e-01
[31] 2.890831e-01 8.359238e-01 5.609066e-01 8.606896e-01 1.648951e-01
[36] 3.132766e-01 1.301822e-01 4.886790e-01 6.948110e-01 9.698405e-01
[41] 5.810904e-01 **2.093289e-02** 1.919763e-01 5.571031e-01 8.082767e-01
[46] 9.216217e-01 6.961176e-01 5.203076e-01 7.875816e-01 8.305888e-01
[51] 5.076901e-01 4.063713e-01 **3.188679e-02** 2.902242e-01 1.317194e-01
[56] 4.913088e-01 7.938832e-01 8.648611e-01 7.551058e-02 7.521584e-01
[61] 4.307146e-01 9.643699e-01 7.331071e-01 3.429180e-01 2.573583e-01
[66] 8.496360e-01 8.375907e-02 6.025290e-02 5.072368e-01 7.350203e-01
[71] 6.604462e-01 5.748327e-01 8.384319e-01 6.925151e-01 9.218235e-01
[76] 8.306421e-01 9.028943e-01 5.518729e-01 1.415273e-01 **4.175693e-02**
[81] 9.692044e-01 6.953160e-01 5.741842e-01 **2.355928e-02** 6.738139e-01
[86] 6.543856e-01 9.223448e-01 7.987887e-01 7.937079e-01 5.326711e-02
[91] 1.421506e-01 6.581521e-01 9.448746e-01 3.545114e-01 9.906047e-01
[96] 1.141325e-01 1.193105e-01 6.938444e-01 **1.790798e-02** 5.482133e-01
```

After the Bonferroni correction is applied, only the first two genes have low *p*-values.

```
> pvaluesBonferroni = p.adjust(pvalues, method = "bonferroni", n =
length(pvalues))
> pvaluesBonferroni
 [1] 5.415589e-20 4.529549e-03 1.000000e+00 1.000000e+00 1.000000e+00
 [6] 1.000000e+00 1.000000e+00 1.000000e+00 1.000000e+00 1.000000e+00
[11] 1.000000e+00 1.000000e+00 1.000000e+00 1.000000e+00 1.000000e+00
```

```
[16]  1.000000e+00  1.000000e+00  1.000000e+00  1.000000e+00  1.000000e+00
[21]  1.000000e+00  1.000000e+00  1.000000e+00  1.000000e+00  1.000000e+00
[26]  1.000000e+00  1.000000e+00  1.000000e+00  1.000000e+00  1.000000e+00
[31]  1.000000e+00  1.000000e+00  1.000000e+00  1.000000e+00  1.000000e+00
[36]  1.000000e+00  1.000000e+00  1.000000e+00  1.000000e+00  1.000000e+00
[41]  1.000000e+00  1.000000e+00  1.000000e+00  1.000000e+00  1.000000e+00
[46]  1.000000e+00  1.000000e+00  1.000000e+00  1.000000e+00  1.000000e+00
[51]  1.000000e+00  1.000000e+00  1.000000e+00  1.000000e+00  1.000000e+00
[56]  1.000000e+00  1.000000e+00  1.000000e+00  1.000000e+00  1.000000e+00
[61]  1.000000e+00  1.000000e+00  1.000000e+00  1.000000e+00  1.000000e+00
[66]  1.000000e+00  1.000000e+00  1.000000e+00  1.000000e+00  1.000000e+00
[71]  1.000000e+00  1.000000e+00  1.000000e+00  1.000000e+00  1.000000e+00
[76]  1.000000e+00  1.000000e+00  1.000000e+00  1.000000e+00  1.000000e+00
[81]  1.000000e+00  1.000000e+00  1.000000e+00  1.000000e+00  1.000000e+00
[86]  1.000000e+00  1.000000e+00  1.000000e+00  1.000000e+00  1.000000e+00
[91]  1.000000e+00  1.000000e+00  1.000000e+00  1.000000e+00  1.000000e+00
[96]  1.000000e+00  1.000000e+00  1.000000e+00  1.000000e+00  1.000000e+00
```

The data for the third gene are now changed to have a mean value of 0.3 to make it more difficult to distinguish between random fluctuations and a genuine difference.

```
> differentGene3 = array(rnorm(50,0.3,1))
> Exp1with50subjects100genes[,3] = differentGene3
> mean(Exp2with50subjects100genes[,3])
[1] -0.2552059
> mean(Exp1with50subjects100genes[,3])
[1] 0.4428331
> for (i in 1:100)
+ pvalues[i] = t.test(Exp1with50subjects100genes[,i],Exp2with50subjects
100genes[,i])$p.value
> pvalues
 [1]  5.415589e-22  4.529549e-05  1.294005e-03  1.486974e-02  9.054391e-01
 [6]  7.098275e-01  4.704718e-01  8.877928e-01  9.196552e-01  9.164268e-01
[11]  1.915345e-01  8.739235e-01  1.242989e-01  9.936854e-01  7.761322e-01
[16]  6.143852e-01  2.527180e-01  4.983262e-01  5.364489e-01  3.473893e-01
[21]  7.880777e-02  3.001068e-01  9.102016e-01  4.904164e-01  3.241277e-01
[26]  6.777727e-02  1.861560e-01  9.631817e-01  8.152621e-01  9.847419e-01
[31]  2.890831e-01  8.359238e-01  5.609066e-01  8.606896e-01  1.648951e-01
[36]  3.132766e-01  1.301822e-01  4.886790e-01  6.948110e-01  9.698405e-01
[41]  5.810904e-01  2.093289e-02  1.919763e-01  5.571031e-01  8.082767e-01
[46]  9.216217e-01  6.961176e-01  5.203076e-01  7.875816e-01  8.305888e-01
[51]  5.076901e-01  4.063713e-01  3.188679e-02  2.902242e-01  1.317194e-01
[56]  4.913088e-01  7.938832e-01  8.648611e-01  7.551058e-02  7.521584e-01
[61]  4.307146e-01  9.643699e-01  7.331071e-01  3.429180e-01  2.573583e-01
[66]  8.496360e-01  8.375907e-02  6.025290e-02  5.072368e-01  7.350203e-01
[71]  6.604462e-01  5.748327e-01  8.384319e-01  6.925151e-01  9.218235e-01
[76]  8.306421e-01  9.028943e-01  5.518729e-01  1.415273e-01  4.175693e-02
[81]  9.692044e-01  6.953160e-01  5.741842e-01  2.355928e-02  6.738139e-01
[86]  6.543856e-01  9.223448e-01  7.987887e-01  7.937079e-01  5.326711e-02
[91]  1.421506e-01  6.581521e-01  9.448746e-01  3.545114e-01  9.906047e-01
[96]  1.141325e-01  1.193105e-01  6.938444e-01  1.790798e-02  5.482133e-01
```

This time, Bonferroni tends to overcorrect to the extent that the significant difference in the third gene is missed.

```
> pvaluesBonferroni = p.adjust(pvalues, method = "bonferroni", n =
length(pvalues))
> pvaluesBonferroni
 [1] 5.415589e-20 4.529549e-03 1.294005e-01 1.000000e+00 1.000000e+00
 [6] 1.000000e+00 1.000000e+00 1.000000e+00 1.000000e+00 1.000000e+00
[11] 1.000000e+00 1.000000e+00 1.000000e+00 1.000000e+00 1.000000e+00
[16] 1.000000e+00 1.000000e+00 1.000000e+00 1.000000e+00 1.000000e+00
[21] 1.000000e+00 1.000000e+00 1.000000e+00 1.000000e+00 1.000000e+00
[26] 1.000000e+00 1.000000e+00 1.000000e+00 1.000000e+00 1.000000e+00
[31] 1.000000e+00 1.000000e+00 1.000000e+00 1.000000e+00 1.000000e+00
[36] 1.000000e+00 1.000000e+00 1.000000e+00 1.000000e+00 1.000000e+00
[41] 1.000000e+00 1.000000e+00 1.000000e+00 1.000000e+00 1.000000e+00
[46] 1.000000e+00 1.000000e+00 1.000000e+00 1.000000e+00 1.000000e+00
[51] 1.000000e+00 1.000000e+00 1.000000e+00 1.000000e+00 1.000000e+00
[56] 1.000000e+00 1.000000e+00 1.000000e+00 1.000000e+00 1.000000e+00
[61] 1.000000e+00 1.000000e+00 1.000000e+00 1.000000e+00 1.000000e+00
[66] 1.000000e+00 1.000000e+00 1.000000e+00 1.000000e+00 1.000000e+00
[71] 1.000000e+00 1.000000e+00 1.000000e+00 1.000000e+00 1.000000e+00
[76] 1.000000e+00 1.000000e+00 1.000000e+00 1.000000e+00 1.000000e+00
[81] 1.000000e+00 1.000000e+00 1.000000e+00 1.000000e+00 1.000000e+00
[86] 1.000000e+00 1.000000e+00 1.000000e+00 1.000000e+00 1.000000e+00
[91] 1.000000e+00 1.000000e+00 1.000000e+00 1.000000e+00 1.000000e+00
[96] 1.000000e+00 1.000000e+00 1.000000e+00 1.000000e+00 1.000000e+00
```

In contrast, if the Benjamini–Hochberg method is used, the first three genes are the only ones left with p-values lower than 0.5.

```
> pvaluesBH = p.adjust(pvalues, method = "BH", n = length(pvalues))
> pvaluesBH
 [1] 5.415589e-20 2.264774e-03 4.313350e-02 3.365612e-01 9.936854e-01
 [6] 9.936854e-01 9.936854e-01 9.936854e-01 9.936854e-01 9.936854e-01
[11] 7.383704e-01 9.936854e-01 6.461390e-01 9.936854e-01 9.936854e-01
[16] 9.936854e-01 9.191367e-01 9.936854e-01 9.936854e-01 9.847538e-01
[21] 5.583938e-01 9.680864e-01 9.936854e-01 9.936854e-01 9.822052e-01
[26] 5.583938e-01 7.383704e-01 9.936854e-01 9.936854e-01 9.936854e-01
[31] 9.674139e-01 9.936854e-01 9.936854e-01 9.936854e-01 7.169352e-01
[36] 9.789893e-01 6.461390e-01 9.936854e-01 9.936854e-01 9.936854e-01
[41] 9.936854e-01 3.365612e-01 7.383704e-01 9.936854e-01 9.936854e-01
[46] 9.936854e-01 9.936854e-01 9.936854e-01 9.936854e-01 9.936854e-01
[51] 9.936854e-01 9.936854e-01 3.985849e-01 9.674139e-01 6.461390e-01
[56] 9.936854e-01 9.936854e-01 9.936854e-01 5.583938e-01 9.936854e-01
[61] 9.936854e-01 9.936854e-01 9.936854e-01 9.847538e-01 9.191367e-01
[66] 9.936854e-01 5.583938e-01 5.477536e-01 9.936854e-01 9.936854e-01
[71] 9.936854e-01 9.936854e-01 9.936854e-01 9.936854e-01 9.936854e-01
[76] 9.936854e-01 9.936854e-01 9.936854e-01 6.461390e-01 4.639659e-01
[81] 9.936854e-01 9.936854e-01 9.936854e-01 3.365612e-01 9.936854e-01
[86] 9.936854e-01 9.936854e-01 9.936854e-01 9.936854e-01 5.326711e-01
[91] 6.461390e-01 9.936854e-01 9.936854e-01 9.847538e-01 9.936854e-01
[96] 6.461390e-01 6.461390e-01 9.936854e-01 3.365612e-01 9.936854e-01
```

REFERENCES

Benjamini, Y., and Hochberg, Y. Controlling the false discovery rate: A practical and powerful approach to multiple testing. *Journal of the Royal Statistical Society Series B* **57**, 289–300, 1995.

Bonferroni, C.E. Teoria statistica delle classi e calcolo delle probabilità. *Pubblicazioni del R Istituto Superiore di Scienze Economiche e Commerciali di Firenze* **8**, 3–62, 1936.

Hoerl, A.E. and Kennard, R.W. Ridge Regression: Biased estimation for nonorthogonal problems. *Technometrics* **12**(1), 55–67, 1970.

Tibshirani, R. Regression shrinkage and selection via the Lasso. *Journal of the Royal Statistical Society. Series B (Methodological)* **58**(1), 267–288, 1996.

Index

Milton Keynes UK
Ingram Content Group UK Ltd.
UKHW040446071024
449327UK00020B/1024